REPETITIVE
MANUFACTURING
PRODUCTION
PLANNING

REPETITIVE MANUFACTURING PRODUCTION PLANNING

Robert A. Gessner

WILEY

A WILEY-INTERSCIENCE PUBLICATION
JOHN WILEY & SONS
New York • Chichester • Brisbane • Toronto • Singapore

Library of Congress Cataloging-in-Publication Data

Gessner, Robert A.
  Repetitive manufacturing production planning.

  "A Wiley-Interscience publication."
  1. Repetitive manufacturing systems. 2. Production
planning. I. Title.
TS176.G497 1987   658.5′03   87-21067
ISBN 0-471-84836-0

Printed in the United States of America

10 9 8 7 6 5 4 3 2 1

PREFACE

This book addresses repetitive manufacturing, both from a systems planning standpoint and with approaches to improve operations.

If you are a producer who has only a few products, you don't really need complex, long-range, sophisticated planning systems. You do need to seriously consider the cost reduction actions that are described.

As a producer of a wide variety of products, if you do not incorporate both the refined planning system approaches and the cost reduction actions, your product lines could be eaten away until only your least profitable items remain and you are out of business.

The job shop producers must recognize that regardless of how long they have produced products, a newer and better way exists, and if they do not adapt, they will no longer be competitive.

This book provides a combination of common sense logic and a case study example for the changes that must be incorporated in the planning systems for a mixed job-shop/flow-shop production environment.

ROBERT A. GESSNER

Kennesaw, Georgia
August 1987

CONTENTS

REPETITIVE MANUFACTURING PRODUCTION PLANNING

PART ONE

REPETITIVE MANUFACTURING AND WHERE IT APPLIES

CHAPTER ONE

IDENTIFICATION OF THE REPETITIVE FLOW SHOP

THE BASIC MEANING OF REPETITIVE MANUFACTURING

The dictionary definition of repetitive is, "Characterized by repetition; tending to repeat." By this definition, if a contract is awarded to built two aircraft carriers over five years, then the production facility is obviously a repetitive facility. This is not the definition that is used in this book. In our discussion "repetitive" means making the same (not just similar) thing over and over. The aircraft carriers would be similar. Due to the incorporation of technological advances over the 5 years, the two vessels would not be identical. The second one would be more state-of-the-art than the first one.

When we discuss repetitive manufacturing, we normally talk about making the identical thing over and over during a rather short time frame. The time may be minutes, days, or weeks. The product, however, is the same thing with the possibility of some applied engineering changes.

By the nature of the above definition, it is also possible to imply product cost in the definition of repetitive. In most cases it is difficult to think of a product costing $100,000 to be built in a relatively short time frame with every successive product identical to the original. On the other hand, it is easy to think of the production of a stereo headset at a cost of $12.49 to be repetitive.

Repetitive manufacturing, by today's industry definition, means the production of a product that has a short manufacturing time, a low cost, and a reasonably high volume.

Recognize that this definition is subjective, which often causes confusion. A company makes mud pumps for oil drilling rigs. The contract is for 500 pumps

3

at $6000 each. Not counting material lead times, it takes two weeks to build one pump. This company would consider itself a repetitive manufacturer. Another company makes 15,000 kitchen clocks a day at a cost of $4.78 each. This company also considers itself to be repetitive. They are both right. They are both repetitive.

What if a company has a basic product line but also offers options and or variants? The cataloged product comes in black, but it can be ordered in red or blue. It can also be equipped with or without a radio. As a result, all of the combinations shown in Table 1.1 are possible. It appears that the one product is actually six different products. The real question is not how many models, options, or features exist, but when in the manufacturing cycle are they installed. In the above example, painting the product might be the last step that is performed. The radio might be added as a dealer option. The major work content in this product is repetitive.

CLASSIFICATION OF REPETITIVE FLOW SHOPS

One way to classify the industry is shown in Figure 1.1.

A SHORT DEFINITION OF TERMS

The following terms are used in the preceding classification.

Shop Order. This is usually a document that specifies that a certain quantity (the lot size) of a particular part is to be produced. The lot size (for example, the requirement to make 500) is normally based on the cost (time) to set up the machine for production of the part and the costs to store the parts that do not have an immediate customer demand. The shop order usually is assigned a unique shop order number and includes desired start and stop dates, plus an indication of the resources (work centers and machines) to be used.

TABLE 1.1   An Options List

Combination	Color	Radio
1	Black	No
2	Black	Yes
3	Red	No
4	Red	Yes
5	Blue	No
6	Blue	Yes

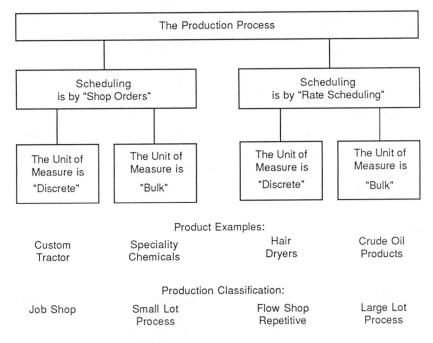

FIGURE 1.1 Classification of Production Types

Rate Scheduling. In this situation, the plant floor (for example) is told to make 1000 per day of a specific item for 15 days. Machine setups are usually not a consideration (although they may be when multiple products are made on the same line). Rate scheduling logic will vary based on the production techniques and the company's policy regarding labor.

A rate may be fixed because the process requires a 500 gallon mixing tank for eight hours. This constraint fixes the rate at 500 gallons per shift.

A rate may be fixed because a transfer line or a belt operates at a certain speed. A transfer line is an automated flow line in which specific machines perform specific functions and movement from one machine to another is controlled by a transfer machine.

A rate may be fixed due to a company's policy on labor. "We have a steady employment policy for our employees." This means that at times product will be produced for which is no immediate customer demand. This also implies that some accurate systems are in place to predict how much of what product is going to be needed during peak demand periods.

Many small companies operate on a variable labor system. They will hire when customer demand is high and lay off when customer demand is low. In this situation, there are often three rates that can be applied to the product of any product.

A minimum rate, which is the slowest the line can run and still be cost effective. This assumes that required machines will still have to be available (at a slower rate), and work stations will still have to be supported by labor (at a slower pace, but at the same pay). The minimum rate is usually 5 to 15% below the desired rate.

A desired rate, which is frequently the rate that top management has decided they would like to operate with whenever possible. This rate can often be exceeded to accommodate peak customer demand periods.

A maximum rate, which is usually limited by the process. The belt does not run any faster. The maximum rate is often 5 to 15% above the desired rate.

These rate scheduling considerations may be summarized as shown in Table 1.2.

Discrete Unit of Measure. This is an "each." It is a single identifiable thing. You cannot have 9/10 of an each. You either have the discrete item or you do not have it.

Bulk Unit of Measure. This is the nondiscrete or continuous unit of measure. It can include fractions of one such as, one and a half gallons of a fluid, the six feet and three inches of an item, or the 4.86 cubic feet of another item.

A unit of measure may not be the same throughout the manufacturing process. For example, paint is mixed using bulk units of measure and then it is put into cans. You cannot buy part of a gallon can of paint. You can buy a quart can, and you can buy a pint can. The gallon, quart, and pint cans are all discrete units of measure. The unit of measure can change from one portion of the manufacturing process to another. By definition, the production process (job shop versus flow shop) can also change from one manufacturing step to another.

THE PURPOSE OF THIS BOOK

This chapter has developed a very simple set of distinctions that allow us to classify job shops as unique from repetitive flow shops.

In the production cycle of any end item (the thing that ships to the cus-

TABLE 1.2 Rate Scheduling Considerations

Group	Rate Type	Environment
1	Fixed	Process size constraint
	Fixed	Process speed constraint
	Fixed	Steady employment labor policy
2	Variable	Variable employment labor policy

tomer), components may be made in a job shop using shop order scheduling techniques; low-volume subassemblies may also be made in job shops; high-volume subassemblies (which are used in many different end items and are also sold as service parts) may be made using rate-scheduled flow shop techniques; and lastly, final assembly of the end item may be rate scheduled.

The purpose of this book is to discuss some of the characteristics of repetitive manufacturing and to illustrate some examples of how high-level planning can be performed in an environment where the production process must address a mixture of flow and job shop planning techniques.

The next chapter presents current industry buzz words as a method of comparing Japanese and American approaches. In some cases we copy the Japanese to remain competitive. In other cases, the ideas turn out to be just good common sense.

CHAPTER TWO

UNDERSTANDING REPETITIVE "BUZZ WORDS"

A new wave of often confusing buzz words currently exists in the industry. Is CFM the same as Repetitive? Are they both the same as Just-In-Time? Let's take a look at some of the more common terms.

AUTOMATION

In a manufacturing discussion, "automation" does not refer to the use of computers in the data processing department to do payroll. It does imply the act of accomplishing work without the use of humans. On the shop floor, automation may consist of:

A computer controlled stacker crane.

A tape or computer controlled machine that cuts chips.

A computer controlled machine that changes tools, loads blanks, makes the part, unloads the part, and then sets itself up to make a different part.

An Automated Guided Vehicle (AGV) that delivers parts throughout the plant.

A quality control sensor system that shuts down the process when tolerances vary beyond required specifications.

However, "automation" in its broadest sense means the assistance provided to humans by computers. A plant that produces 300 end items with 20,000 different components needs a Material Requirement Planning (MRP) system. A manual MRP system would be very impractical. It would take days to develop a plan that would be obsolete when it was completed. A computer will perform the task in minutes. Therefore, in addition to the benefits of automation on the shop floor, there are automation benefits in the planning and scheduling areas of the manufacturing process.

BALANCED FLOW

In the United States this phrase normally means only the balanced flow of the production process. The Japanese view "balanced flow" as relating to:

The balanced production process at supplier locations for all required components and materials.

The balanced supply lines that deliver components from the supplier to the plant that produces end items.

The balanced production process of the end items.

The balanced distribution of completed end items to the consumer.

The balanced distribution of service parts necessary to maintain the products that were sold.

Some of the major U.S. companies are starting to adopt the Japanese view, but most American companies still consider "balanced flow" as something that applies to only the production process.

There are basically two ways to achieve a balanced flow in the production process. In the first case, you know up front the quantity you want to produce. You set up the plant (or line) to produce that quantity in a nice smooth flow with no bottleneck processes. Frequently a simulation tool is used for this initial plant setup operation.

In the second case, what you produce each day at what quantity is known to be variable. On a daily basis (or some short period of time), you may want to be able to readjust which tasks are performed at what work stations (based on how many people showed up for work that day). This job is often accomplished with a line balancing tool which calculates a best case solution based on the number of available workers, the amount of product to be produced, the number of available workstations, and the network relationship of the tasks to be performed to produce products.

Note that the term "balanced flow" does not apply to a job shop. Job shops are plagued with imbalances due to lot sizing criteria that varies from one work center to another and shop order schedule priorities. Job shops are designed to produce many different end items using multipurpose machines. Repetitive flow shops are designed to produce one (or several similar) end item in a continuous flow with single purpose machines that do not require setup changes.

COMPUTER INTEGRATED MANUFACTURING (CIM)

Computer Integrated Manufacturing (CIM) is the design goal of many major manufacturing companies for the 1990s. There are three major segments of a manufacturing business that comprise a CIM system.

Production and Business Systems (P&BS)
Engineering Systems
Plant Floor Operation Systems

In the early 1980s, although these systems existed, the interfaces between them were nonexistent, manual, or poor at best. Each of these system areas were developed by different people (different backgrounds in different departments) and each group wanted to get their own job done. Little thought was given to how the people in other departments can do their job better by having access to the data that I have developed. Computer Integrated Manufacturing represents the tight integration of these three basic system areas.

In addition to these three major areas, there are three other areas that are frequently considered to be part of CIM.

Financial Systems
Office Systems
Electronic Data Interchange (EDI) Systems

This book deals with only a part of the P&BS portion of CIM. Production and Business Systems covers the following functional areas, but this book addresses only those areas marked with an asterisk in any real degree of depth.

Forecasting
Customer order servicing
Master (family level) production planning*
Resource requirements planning*
Master schedule planning*
Rough-cut capacity planning*
Bill of material processing
Inventory accounting
Material requirements planning
Purchasing
Receiving
Capacity requirements planning
Shop order release
Facilities data management

Routings
Product costing
Cost accounting

CONTINUOUS FLOW MANUFACTURING (CFM)

There appear to be several meanings to this term, depending on the type of business to which it is applied. In most cases, it is used interchangeably with the term "repetitive." A repetitive manufacturer of toy trucks will often state that they employ Continuous Flow Manufacturing (CFM). They make many different types of discrete items, one shift a day.

An oil refinery claims that they do CFM. They produce bulk products 24 hours a day, 7 days a week.

A textile firm produces a continuous flow of a basic cloth. The plant runs at two shifts per day, five days a week. The product, measured in square feet, is a bulk product. Since the flow of the cloth is continuous (until cut and shipped as bolts), this firm says they employ in CFM.

An electronics firm sets up a very sophisticated transfer line to manufacture a computer component. The line will run at one shift a day for several weeks before it is modified to make a different component. This firm will claim to perform CFM.

And, of course, there are those unique definitions. One company stated that if they produced the product for stock, it was CFM. If they produced the same product for a large customer order (as 1000 per day for 15 days), it was repetitive.

In general, we can assume that when dealing with discrete items, CFM has the same meaning as repetitive.

DAILY MACHINE CHECK

This is a concept that comes from Japan and ties in with "diversified skills." In the United States, we have a skill group that operates machines and another that maintains machines. In Japan, operators are not only responsible for making high-quality products, they are also responsible for the basic maintenance of their machines and for the maintenance of their respective work areas.

Although this concept initially met with some resistance in the United States, it is now being understood that for companies to reduce costs, remain competitive, and keep their direct labor employed, it is one of the new ways of thinking which has to be implemented.

DIVERSIFIED SKILLS

Ask a U.S. worker what he does for a living. You will probably get an answer like, "I run a 25 spindle omni mill. I'm a specialist. I've run this same mill for 15 years." You ask, "If that's the only machine you work on, what do you do when there is no work for your machine?" In many cases, the honest answer is that the operator works slower.

The Japanese recognized this problem. They saw an opportunity to increase productivity while expanding the horizons of their direct labor personnel. Awards and promotions were based on an operator's ability to run multiple machines. This has increased the ability to shift personnel from slack areas to those facilities that are currently bottlenecks.

Acceptance of this concept has been slow in the United States. While it is easy to say that it is obviously a union problem, it is really more of a personnel problem. The milling machine operator has been doing one thing for a long time. He is good at it. He has status. Now you are going to ask him to learn three other machines. He may not be quite as good as those. He may lose some status among his peers.

The solution may be education. It may be the type of motivation that is employed. It might be attrition and just take time as the old way of thinking finally retires.

GROUP TECHNOLOGY

This is an item classification system. As of this writing, there is no standard on the classification technique. The concept evolved in different companies which used different classification approaches, depending on their needs. Classifications are sometimes based on the following different approaches:

By the function that the items are to perform when they are used

By the raw material used in the items

By the manufacturing steps required to produce the items

By the shape of the items

The group technology technique is very useful in a manufacturing company. Retrievals of classified items can be used by:

A design engineer who wants to design an item to perform a function similar to the function performed by a previously designed item

A manufacturing engineer who wants to find the correct alloy for the item which will satisfy design requirements and yet can easily be produced

An industrial engineer who wants to know how to set up the line based on how the manufacturing steps were set up for a previous similar product

A materials planner who wants to see if this item has a similar form, fit, and function to a previously designed (and currently in stock) item

Standards must be agreed upon for group technology before general purpose systems will be available for every manufacturer, regardless of the product that is produced.

HOUSEKEEPING

In America, the following skills or trades are usually present at a work center.

Operators
Setup personnel
Machine maintenance personnel
Janitorial personnel

The Japanese maintain that keeping an immediate work area clean should not be the function of external personnel like a janitorial service. It should, in fact, be the responsibility of the person who gets it dirty. Machine operators will tend to be more careful about letting their work area become dirty, since not only will they have to clean it, but they will also be measured on how well the area is maintained.

Reluctance to accept this concept in America appears to be based on operators feeling that they are skilled craftsman (which they are), and they are above doing such a menial job as pushing a broom.

JUST-IN-TIME (JIT)

You can build it Just-In-Time (JIT) to make a shipment (Japanese approach), or you can make it Just-In-Case you might ever need to make a shipment (American approach). This simple (and not wholly accurate) analogy indicates why JIT is frequently associated with low and even zero inventories.

When the JIT concept was first introduced in the United States, many people viewed it as the solution to all manufacturing problems. Planning systems would no longer be required. Material Requirements Planning (MRP) could be eliminated and all of those high inventories would be reduced to zero. Shop floor reporting was not really necessary, since once a day you notified someone that you just made 10,000 of the product.

If you are in the business of making 10,000 per day of the product, the assumption is correct that all of those sophisticated planning systems can be eliminated. In fact, if your business were that simplistic, it would be questionable why you ever had any sophisticated planning systems at all.

Do not misunderstand. The JIT concept is a good one and will benefit the United States manufacturing industry. It has, however, as many flavors as there are shades of color for a rose. It is probably most totally in control (and effective) in the simple environment. Make just a few products in a labor intensive, assemble only, multiple-shipments-per-day environment. Your lead time is in minutes or hours. Just-In-Time will function very well in a pure pull mode. You do not make anything until you have a need to ship it. When you need to complete a final assembly, you pull the subassemblies from the previous work stations. The subassembly stations now have the need to produce, and they in turn pull sub subassemblies from prior workstations. The pull process goes all the way back to the suppliers for components.

A pure push (non-JIT) system does a lot of planning around some combination of forecasted demand and actual customer orders, and develops a production plan that it pushes through the manufacturing cycle. If demands did not change and all lead times were short, a push system would work as well as a pull system.

However, life is not quite that simple. Manufacturing lead times are sometimes long. A supplier who is 1000 miles away may not be able to respond to your component requirements as they vary every four hours. As a result, many companies are planning to use some combination of push and pull systems.

Just-In-Time is now being considered as a shop floor material replenishment system. There are other attributes, such as the following, which were once considered to be part of JIT.

Have operators do housekeeping
Provide 100% quality
Reduce setups
Make the lot size close to one
Balance the work and material flows
Diversify labor skills
Emphasize preventive maintenance
Control workload flow by visible indicators
Arrange the plant to produce specific products
Build supplier networks
Motivate workers to be involved in quality and production techniques

These attributes are now considered to be just good common sense, but they are still prerequisites to a successful JIT implementation.

KANBAN

This started out as a Japanese technique to pull material. It was a manual approach using cards. When I have used the subassembly that you made for me, I send the card back to you which tells you to make another subassembly. You do not make one unless you have a card. The amount of Work-In-Process (WIP) can therefore be controlled by the number of cards that exist.

Kanban now has many forms. Put identification, location, and quantity information on a tote, tray, or basket, and it then takes the place of the card. The empty container is returned, indicating the need for a replacement.

Electronic signaling devices are also used to say, "I used this one. Make me another one."

Kanban, regardless of the type that is employed, is the basis for the material pull concept of JIT.

MANUFACTURING CELLS

This is a concept that is in between the pure work center and the pure flow-line approaches. It consists of putting together a group of functions (labor and or machine) that can be utilized to produce a similar set of products. From the labor standpoint, it is an extension to diversified skills. Manufacturing cells are used to produce a mix of similar products in as close to a repetitive environment as possible.

QUALITY AT THE SOURCE

I have the first station on this assembly line. You have the second station. You cannot do your thing until I do mine. Each of us is measured on the amount of quality product that we produce. I had a good time last night, but I do not feel too good today. I do a sloppy job on my piece of the assembly. As a result, you have to take the one that I just finished (which is defective), and cannot pick another from a lot of 500 sitting on a pallet, since the pallet does not exist in a repetitive environment. You are not too happy, as I am cutting into your paycheck.

One defective item has been produced, and the defect has been identified at the very next workstation. Result—one item was made defective, and the cause was corrected immediately (you suggested to me that unless I straighten up, I would have to feed your six kids).

In a job shop, I would have made 500 of the item, put them on a pallet and sent them to you to work on. You find that 350 of the 500 items are defective. The 150 good items that I made do, however, give you enough work to do for the last few hours today. In fact, today you only complete 125 of the 150 good items. Tomorrow, you finish the other 25, and report the job as completed with 150 good and 350 as scrap (or requiring rework). The system is updated once a day. The company has now lost two days to find out that they have a defect problem. On the third day, when a supervisor is trying to find the source of the problem, I won't remember or I'll be on vacation or sick leave. The problem frequently never does get solved.

Quality at the source means that when the first defect happens, fix it. Correct the person or machine now. Do not wait until some large quantity of the item has been produced.

QUALITY CIRCLES

This is an employee involvement program. Rather than having just management and engineers say what to build and how, why not let the people who are building the product make improvement suggestions? Scheduled meetings of the direct labor employees are set up to discuss how quality can be improved and costs reduced.

This has become an accepted United States procedure, and in many companies has been very successful.

REPETITIVE

To make more than one of the same item; to repeat. The balance of this book is devoted to describing flow line repetitive manufacturing.

ROBOTICS

When most people think of a robot, they visualize a metal creature, shaped sort of like a human, that came from outer space. The shop floor robot does not normally fit the movie image. It is usually designed to perform one or more specific functions. Modern robots are controlled by computer programs that are usually resident in the robot's controller. The controller may be a stand-alone device, or may be interfaced to a host computer system that can download different computer programs as it becomes necessary to produce different parts.

Robots can be installed to perform many of the functions that are performed by humans today. Once programmed to perform a specific task to produce a quality product, a robot will perform that task continuously with little variation in the quality of the products being produced.

In the early 1980s, many companies viewed the installation of robots as a capital expenditure that could not be afforded. Today these very same companies have realized that they cannot afford not to utilize robotics.

There are many good books on the subject of robotics on the market today, and I will not go into detail on the subject.

STOCKLESS PRODUCTION

This is producing product without excess stock. Consider a simplistic point of view with only two types of stock, input and output. Input stock is the raw materials, components, and subassemblies required to produce the product. Output stock is the finished goods ready for shipment.

In an "ideal" environment, if I wanted to make an X out of an A and a B, I would not have the A and B shipped to me until I needed to make the X. And, as soon as I made the X, I would ship it. As a result, I would remain as close to a zero inventory position as possible.

The "ideal" environment is obviously not practical for most companies. The opposite extreme is that I can "get a good deal" if I order a year's supply of As and Bs. My inventory is high. My inventory carrying costs are high.

Companies today are trying to approach the ideal environment within practical limitations. They give their suppliers a blanket purchase order for the number of components that they think they will use during the year, but they specify the delivery shipments to be weekly or even daily. The supplier has a reasonable estimate of what his production requirements will be for the year, and the manufacturer gains the benefit of some discounts without high carrying costs. One of the trade offs is the cost of the more frequent shipments.

Many companies are aspiring to have suppliers deliver every four hours, and in turn, they will ship to their customers every four hours. When achieved, this will result in what today is referred to as stockless production or zero inventory.

TOTAL QUALITY CONTROL

Quality is what you decide it should be. A soggy, fatty, lukewarm hamburger from a fast food restaurant may be considered to be a quality product, not by you, but by the owner of the restaurant. The owner set standards (specifications) such as:

Use two ounces of highly flavored sauce to cover the taste of the inexpensive meat

Use meat that has not more than a 30% fat content

Prepare the hamburgers at a steady (highly productive) rate and store them under heat lamps for the peak demand periods

The standards were set to keep costs low to address a specific type of potential customer (market). If sales and profit objectives are met, the owner is satisfied with the "quality" product.

The Japanese have determined that quality is better determined by the consumer. If you make the quality that all customers want, you will have more customers. To insure that all customers are getting what they want, all of the products must be at the same quality level. This means that periodic or sampling inspections are no longer acceptable.

Quality can be achieved by two methods. It can be built in, or it can be inspected in. The Japanese prefer to build in quality (see "Quality at the Source"). When they do need to inspect quality in, they advocate doing it on a 100% basis of all products that ship.

The key to total quality control, which many U.S. manufacturing companies have not recognized, is to maximize building quality into the product therefore reducing the need for the 100% inspections to a minimum.

VENDOR NETWORKS

One of the Japanese approaches regarding vendors or suppliers is to build long-term contractual relationships with a minimum number of vendors. In the United States, many companies still have many suppliers for any one component, with whom they constantly renegotiate for lowest price or practice a "spread the business" philosophy. The Japanese approach is now being recognized as a more cost effective approach.

Still, a lot of vendors may continue to be required to produce an end item if many components are involved. And, to keep inventory down, the manufacturer wants daily deliveries from each of those vendors. It would be impractical to expect each vendor to load one box of parts on a truck and ship it to the manufacturer.

A more practical approach is to have a truck (frequently managed by the manufacturer) perform a route pickup service. One truck daily covers a geographic grouping of vendors and picks up the components that are required for the next day's production. The truck can also drop off the pallets and totes that were used for delivery purposes on the previous day.

This approach of developing vendor networks is becoming popular in the United States, especially since many manufacturers are providing incentives for suppliers to move closer to their plants.

ZERO DEFECTS

See "Quality at the Source" and "Total Quality Control."

ZERO INVENTORY

See "Stockless Production."

This has been a short review of some of the concepts that are currently being considered in the United States. The next chapter covers some of the trends in manufacturing that are taking place.

CHAPTER THREE

THE TRENDS OF U.S. MANUFACTURING

RETHINKING THE MANUFACTURING PROCESS

A company in Smalltown, USA started up many years ago to make product A. They purchased some used machines and tools at that time, and started making As. As time went by, they could see a need for Bs, Cs and Ds, and added a few more machines and expanded their product line. Time went by. They now produce 300 different products. They have many machines, which are now grouped by the functions that they perform, such as all drill presses being in a single group (called a work center). The company in this scenario is a job shop. A major percentage of American companies evolved in a manner similar to this scenario and continue to exist as job shops today.

Two things are happening that are causing manufacturing companies to rethink the way they produce products.

Direct labor is decreasing in the manufacturing process due to the technological advancements in automation. Machines will work three shifts a day, they will not ask for overtime, they will not go on strike, they do not take coffee breaks, and, most significantly, they are consistent. Once a machine is set up to produce a quality product, it will produce to the same level of quality by the thousands.

Offshore competition is forcing American companies to reduce costs or become noncompetitive and go out of business. A Japanese firm might review the following situation.

Product A has been on the market for one year.

It is a product only manufactured and distributed in the United States.

It sells at about 500 per month for $200 each.

It has a defect rate of about 5%.

It costs $150 to make an A.

They would ask themselves the following questions:

It we reduced the selling price to $125, how much more of a market could we attract?

If we reduced the defect rate to less than 1%, how much more of a market would we attract?

If we expanded the market to other countries, such as Australia, how much more of a market would we attract?

When these questions were answered, a business case would be developed to ascertain if it would be profitable to set up a facility to produce As. The business case would consider potential production costs for a high-volume repetitive production environment.

Note that the American company may never have considered As to be a high-volume product. They evolved into making As. They also make 400 other items. The Japanese are not planning to make 400 other items at their facility, only As at 2000 per day, at a cost of $98 each to be sold for $125 each worldwide.

These pressures will force many U.S. companies to rethink the way that they produce products if they want to stay in business.

THE SHIFT FROM JOB SHOPS TO REPETITIVE FLOW SHOPS

The emphasis on repetitive flow manufacturing, along with the competitive pressures, will cause many U.S. discrete-product manufacturing companies to migrate from pure job shops toward becoming repetitive flow shops.

The migration will be slow in most cases, with a company selecting a key bottleneck or high-volume production process, converting it to a repetitive process, refining the problem areas out of it, and then selecting the next key conversion area.

Of those companies that do start to migrate from job shops toward repetitive flow shops, most of them will never become 100% repetitive. It will continue to be practical to manufacture some items in a job shop environment. As a result, most future manufacturing companies will produce products in a mixed job-shop/flow-shop production environment.

The following sections illustrate a simple example, constructed to show how master level planning logic can be applied to this mixed environment. A basic plant facility has been defined, consisting of five flow lines and four work centers. Production is planned for nine end items that have been grouped into four product families.

The example provides for the situation where any component at any product structure level of the nine end items may be scheduled with a shop-order lot-sized quantity at a work center or with a rate at a flow line.

PART TWO

FUNCTIONAL CONSIDERATIONS

CHAPTER FOUR

BASIC FUNCTIONS

This section addresses the mainstream of the functional areas in a repetitive environment. It is provided to obtain an overview of the sequential stream of logic that is utilized in scheduling a flow shop.

FORECASTING

If the company has distribution centers, this function is performed at the distribution centers as well as at the plant. The basic input is the historical demands that have occurred at each geographic location (as at a distribution center). This is normally accomplished by aggregating all item demands into sales families and then forecasting the sales families, which allows for forecasting a few families instead of hundreds of items.

Once the initial intrinsic forecast has been developed, it is normally reviewed by business planning, and extrinsic factors are applied to the sales families such as marketing sales objectives, product phase-in and phase-out plans, product cost and price objectives, projected technology changes, and industry trends.

The resulting sales-family forecasts are then disaggregated based on item mix percentages within the family and all common item forecasts (for instance, an item that is sold by multiple distribution centers) are consolidated.

The item forecasts are provided to production planning as an input.

FACILITY SIMULATION

When line balancing is run based on the desired Daily Going Rate (DGR), it provides information as to how the work stations on the line are to be set up and manned. There are, however, many supporting functions that need to be addressed, and this is the function of facility simulation.

Facility simulation provides the balance from a plant viewpoint just as line balancing does from a line viewpoint. It addresses all of the material and labor support functions necessary to keep the balanced line running.

The major components of a manufacturing system are parts, work stations, transporters, conveyors, operators, and fixtures. Parts may be raw materials, work-in-process, subassemblies, or finished products. Parts should arrive at the system according to a specific production schedule and move through the system according to a specific plan or route. In this simulation approach, the route (by user definition) should be deterministic, probabilistic, or conditional, based on the system status and part characteristics.

Material handling equipment moves parts between stations in a manufacturing system. Fork lifts, trucks, and other conventional material handling devices are normally represented as transporters. Within this simulation, a representation of conveyor systems should also be available and could include straight line segments, loops, and conveyor networks. The system should accommodate considerations on the operators necessary to run the material handling equipment, such as, whether an operator can perform a single task or multiple tasks. Alternately, are different operators capable of performing the same tasks? Or, can flexible teams of operators work together on some tasks but not on others?

The stations, material handling equipment, and operators may also be constrained to a specific work period or shift.

Comprehensive displays should provide information on system throughput, equipment and personnel utilization, work-in-process, and system status.

A throughput display should provide statistics on the system throughput broken down by part type, station, transporter type, fixture type, and personnel class.

A time measurement display should provide information on how parts spend their time in the system. For each part, a display should illustrate the total time in the system, dividing this time into four categories: processing time, traveling time, station waiting time, and transporter waiting time.

A station inventory display should give a complete summary of preprocess and postprocess inventories at each station, including statistics on the number of parts in the storage areas as well as statistics on the time spent by these parts when they must wait for transport or processing.

PRODUCTION PLANNING

The objective of production planning is to determine if anticipated and actual demands can be met across a long horizon. The output is often used by top management to decide on the addition of facilities, equipment, and floor space. It is the long-range view of what should we make (based on demands) versus what we have the capability to make. As such, planning is performed on a high (product family) level. In addition, production planning provides item production plans to master schedule planning for shorter horizon planning.

The main areas of the production planning function are production planning and resource requirements planning. The function descriptions that address these areas identify their applicability to the job shop, flow shop, or mixed shop environments.

The entire production planning function, because of its scope, must deal with all items that the manufacturer produces, whether they are rate scheduled or shop order scheduled.

Throughout production planning, a toggle capability should exist so that a planner can switch from discrete period quantities to cumulative quantities to costs to sales dollars.

Where work stations are referenced, the station may be a simple station manned by one person, a multiperson station, or a flexible manufacturing cell. All three are to be treated as a single work station from the standpoint of total required station changeover time. Multiple stations make up a line. When a line is switched from one product family to another, line changeover time is applied.

Production engineering may have utilized group technology coding and, as such, this may be an additional consideration for the establishment or definition of a product family.

Tools should be available to assist a user in the development of product families. The most commonly accepted rule for determining which items should belong to which family is—All items that consume about the same amount of about the same resources should be structured into the same family. Consider a manufacturer that has 800 items in the product line that have been structured into 40 families of about 20 items each. Now item number 801 is introduced. Into which of the 40 families does it best fit, or should it become the first item in the 41st family? Tools need to be made available to the planner to allow this type of decision to be easily made.

Aggregation capability needs to be available so that all of the item details can be summarized to a family level.

Family management capability allows for the management of cumulations, provides for build-aheads, and accommodates building to specific inventory levels. For each family, at least three plans should exist: the plan that was developed at the beginning of the year, the plan that is currently in use, and the plan that is currently being developed.

When a plan is developed at a family level, it is necessary to test it against available resources to determine if it is feasible. In a job shop environment, this requires a roll up of resources identified in the routings required to produce a product, extension by the period quantities, and then consolidation of all of the item quantities within the family. In addition, the user should be able to identify unique resources that need to be evaluated which are not in the product routing. These family requirements (to meet the family production plan) are then matched against the available (and projected available) resources.

For the flow shop environment, the resource requirements planning process is much simpler than for a job shop, since it involves matching DGRs (for the family) against the available time on a line.

Assume that a company can sell 300 different products. In this assumption, a single product, a product model, or a product with an option are all counted as different products. Assume that these 300 products may be grouped into approximately 30 product families with about 10 products per family. A product family is normally considered to be a grouping based on the consumption of common resources. In the repetitive sense, the above definition is extended to mean that all products, models, and options which can be built on one line at the same time (intermixed) can be consolidated into one family.

Assume that only one physical line exists on which all 30 families must be scheduled and the family plans appear as shown in Table 4.1.

The DGRs are obtained from business planning and may be provided as shown in Table 4.2.

The fit calculation is as simple as the example in Table 4.3. This example (which does not include line changeover time) indicates that all requirements for the 30 families can be met at the desired DGR in a 22-day planning period.

Normally the problem is a little more complex. The above example assumes that no product has to ship prior to the end of a 22-day planning cycle. If some

TABLE 4.1   An Example of Family Requirements

Family	Required Quantity
1	15,000
2	40,000
:	:
:	:
30	10,000

TABLE 4.2   Example Daily Going Rates

	Daily Going Rates		
Family	Minimum	Desired	Maximum
1	12,750	15,000	17,250
2	12,750	15,000	17,250
:	:	:	:
:	:	:	:
30	1,700	2,000	2,300

TABLE 4.3 Example Fit Calculation

| Family | Required Quantity | Daily Going Rates | | | | | |
| | | Minimum | | Desired | | Maximum | |
		Rate	Days	Rate	Days	Rate	Days
1	15,000	12,750	1.2	15,000	1.0	17,250	0.9
2	40,000	12,750	3.1	15,000	2.7	17,250	2.3
⋮	⋮	⋮	⋮	⋮	⋮	⋮	⋮
30	10,000	1,700	5.9	2,000	5.0	2,300	4.3
Total	194,600		25.3		22.0		18.7

quantity of each of the 300 products had to ship weekly, then the cycle time would be weekly, and the line changeover costs and time would be four times the monthly cycle time.

In addition:

Specific products may be designated to ship on specific days, causing line changeover calculations to change.

The existence of multiple physical lines must be considered.

It may be possible to run a particular family on two or more physical lines.

It may be possible to subcontract a family's production requirements.

The first step of production planning for a rate scheduled item is therefore to see if it will fit within the available resources across a specific horizon for specific (user defined) time buckets.

In the event a potential plan is not feasible, the capability should exist for the production planner to review alternatives, such as what it would cost to subcontract the requirements for a particular family, or if it can be produced on another line which has a low scheduled load but has a higher changeover cost to run this family.

Product family sequencing determines which product family precedes which other one based on line changeover costs and the demand plan for the items in the family.

The objective is to determine the family which should be scheduled first, which second, and so on. Line changeover costs are normally a key consideration, and station changeover costs of products within a family are not applicable. The process assumes that once a line is set up, all of the products within the family can run intermixed on the line.

A scheduling example might be five families that are called A, B, C, D, and E.

For a specific planning period, the actual demand requirements are critical for the adjustment of the sequence. For example, if the demand for the items in families A, B, and C existed in week one and the demand for the items in families D and E existed in week four, it would be ineffective to make products for all five families during every day of the month.

As an example, assume that no product had to ship until the end of the month and the only constraint was line changeover costs.

These costs would impact the sequence in which the families should be run. A simple matrix can be constructed to reflect the line changeover costs, as shown in Table 4.4.

The best solution to this problem from a least-cost basis is the sequence of A, D, B, C, E, A, D, and so on.

The capability must be available to compare various plans of production and demand, such as:

The January family production plan to the current family production plan or the plan that is now being developed (any of the above three in comparison with any other).

Any of the above three plans compared to the master schedule plan (aggregated to the family level), or the line product scheduling plan (aggregated to the family level).

An item production plan to a master schedule plan.

The current demand picture compared to any of the above plans.

MATERIAL REQUIREMENTS SCHEDULING

The objective of this function is to establish the rates and durations (for rate scheduled items) and the order quantities (for shop order scheduled items) for all supporting feeder lines and work centers.

Product scheduling and line balancing established the rates, durations, and sequences for the bottleneck line in the production process. This function is

TABLE 4.4 Family Line Change Over

| From Family | Costs to Change the Line | | | | |
| | To Family | | | | |
	A	B	C	D	E
A	—	60	100	70	50
B	90	—	110	80	30
C	100	65	—	80	40
D	80	70	120	—	50
E	20	75	90	90	—

required to assure that all supporting facilities can provide components in the right quantity and at the right time.

If there were a plant with a network of five flow lines and four work centers that produced three families of products, and three of those facilities were bottleneck facilities, this function would provide the scheduling support to the remaining six facilities (with the assumption that product scheduling addresses the critical facilities).

PRODUCT SCHEDULING

The objective of product scheduling is to determine the sequence and mix quantities of products to run on a line. In production planning, the sequence that product families should be scheduled on a line was established. At this point, the sequence of the products within a family is established, along with the associated quantities for each iteration.

Assume the following:

Product family PF1 is to run on line six.

PF1 contains three products: A, B, and C.

The production plan for the products is shown in Table 4.5.
Based on work station changeover costs, the best sequence to run the products is determined to be B, followed by A, followed by C, followed by B, and so on. (The changeover cost calculation is similar to that used in production planning for product family sequencing)

If shipments are made every five days, the schedule should be to make 1000 Bs, followed by 500 As, followed by 100 Cs, followed by 1000 Bs, and so on.

If shipments are made every four hours, the schedule should be to make 100 Bs, followed by 50 As, followed by 150 Cs, followed by 100 Bs, and so on.

If shipments are made every hour, and work station changeover costs have been reduced to zero, the schedule should be to make 2 Bs, 1 A, and 3 Cs in any sequence that is convenient.

The product scheduling process should also take into consideration the fact that the work content may vary for the above three products, and it may be totally impractical to schedule, for example, 150 Cs at one time as it would

TABLE 4.5   Sample Family Requirements

Product	Daily Planned Quantity
A	100
B	200
C	300

bog down the line. This problem requires the application of both product scheduling and line balancing for an optimum solution.

Product scheduling develops the sequence that products (within a family) should be run on a line, determines the proper run quantities for each iteration, and provides the daily schedules to production release.

BATCH/LOT CONTROL

This function is required by Department of Defense, food and pharmaceutical manufacturers and their suppliers who are monitored by the FDA, USDA, EPA, and by various state agencies. It is also needed by manufacturers of complex electronic products. The capability should exist to:

Track vendors who supply batch/lots.

Track received batch/lots through quality receiving inspection.

Track manufactured batch/lots through the plant.

Track batch/lots through the distributing channels to wholesale and retail customers.

Support is necessary for batch/lot allocation and planning, five position decimal quantity fields and yield, potency, and shelf-life factors.

BAR-CODE USAGE

The objective of this section is to identify the need for the use of bar-code preparation and reading where materials are tracked in the system. In some instances, the bar code may identify a specific container which in turn identifies the material and quantity in the container. In other cases it directly identifies a specific item.

Bar codes are being used to track both materials and containers from suppliers, within the plant, and to customers.

A container of material is shipped from a supplier. It contains a single bar code. A single bar code can reference the following types of data that are used by the supplier, by the shipper, and by the receiver:

Part number

Customer code

Customer name

Supplier code

Supplier name

Customer dock code

Purchase order

Customer contact person

Date issued

Time to receive

Time to ship

Time issued

Quantity required—Cumulative YTD prior to time #1

Cumulative shipped YTD

Last shipment date

Last shipment identification number

Time 1–N: quantity required net

Time 1–N: quantity required cum

Time 1–N: date required

Time 1–N: time required

Record keeping year

Bar-code creation and reading capability is necessary in the industry for tracking purposes. It should be able to be functional in all environments including: the receiving dock, the shop floor, and departmental offices.

PERFORMANCE MEASUREMENTS

The flow shop environment historically manages their data by cumulative quantities. For example:

Production planning: Family cumulative planned versus family cumulative actual.

Material requirements planning: Item cumulative required quantities versus item cumulative produced quantities.

Purchasing: Cumulative measurements against blanket order quantities.

Receiving: Cumulative measurements of shipment notifications against shipment notifications.

These cumulative measurements are not just simple totals, but are displayed by cumulative performance against time as shown by the example in Table 4.6.

As DGRs are by period, the cumulative data is frequently applied to performance data for a daily (and sometimes hourly) measure of how well the lines are performing.

Support is required for all quantity data by specific quantity, by period quantity, and by cumulative quantity.

TABLE 4.6 Cumulative Example

Cumulative Shipped	Change Quantity	Quantity Due 8/22	Cumulative Required 8/22	Cumulative Requirements through			
				Week of 8/22	Week of 8/29	Week of 9/05	Week of 9/12
72,500	1,428	900	74,000	73,400	74,400	75,201	76,199

HISTORICAL DATA ANALYSIS

The objective of this function is to insure that any type of data analysis can be performed. This means that all transactions must be saved in their original form. The situation where transactions are summarized and then discarded with only the summary data retained cannot exist if an adequate data base is to exist for historical data analysis.

CHAPTER FIVE

YIELD MANAGEMENT

BASIC YIELD MANAGEMENT CONCEPTS

The objective of this function is to provide variable yield management controls by work station. The yield trends over time are used as projections for production planning and for current problem analysis. Yield scheduling takes place in both forward and backward modes and provides the data to determine material requirements at each work station, as well as determining the quantity of product that can be produced from a limited amount of material. It also projects material requirements for planning systems based on yield trends.

The two basic scheduling techniques are backward and forward scheduling. Backward scheduling might stipulate that 2000 items constitute a Standard Batch Quantity (SBQ) to be produced. Backward scheduling starts at the last work station and proceeds back to the first work station using the yield factors at each work station to determine the quantity of input materials required at each work station. Backward scheduling, therefore, has an input of the desired production quantity and calculates the input component or material requirements at each work station.

Forward scheduling takes place when the manufacturer identifies a limited amount of an input component or material and wishes to find out how much of the final product can be produced.

Note that if a critical component is required on an intermediate work station, such as work station three, then backward scheduling takes place from station three to station one and forward scheduling takes place from station three to the last station.

Yield calculations should accommodate the entire process, such as by products, coproducts, graded products, and rework loops.

Work station yields are obtained from production reporting. The yield data is used to develop projections over time that are then used in production planning and material requirements planning. Note that yield management also has interfaces to batch/lot control for tracking lots of components and/or lots of the finished products. As such, work station yield management applies to a job shop that releases a work order for a SBQ and a flow shop that schedules based on a daily going rate.

The process to which yield management applies may consist of:

A single task/operation (at a work station)
A sequential string of tasks/operations (at one or more work stations)
A network of tasks/operations

A task/operation consists of the application of resources (labor and or machine) to the production of one or more items. A task/operation may be used to:

Assemble
Disassemble
Separate
Inspect
Test
Change characteristics (i.e., by heat or pressure)
Change shape (i.e., by machining)

The possible task/operation inputs are:

Item identification
Source (the previous task/operation or stock point)
Quantity, as:
    A standard batch quantity
    A lot sized quantity
    An overlapped lot sized quantity (which may range from two to all tasks/operations)
    A tote quantity (which may be similar to a lot sized quantity except that the totes may be serial number controlled)
    Discrete item quantities

Possible operation outputs are:

Item identification
Destination (the next operation or stock point)

Quantity, as:
  Prime product
  Coproduct
  Byproduct
  Trim
  Scrap
  Graded product by:
    Specification
    Potency
  Recovered product
  Reworkable product
Machine time
  Setup
  Run
Labor time
  Setup
  Run

Yield calculations are based on task/operation percentages (where one or more tasks/operations can exist at a work station) and can be performed:

  Forward (for a given amount of material, how much final product can be produced)
  Backward (to produce a desired amount of final product, how much material is required)
  Both forward and backward (where the constrained material is not at the first task/operation or the final product is not at the end of the process)

Four examples of yield calculation logic are included below to illustrate the types of calculations that take place in yield management.

Note that the terms "tasks" and "operations" are used interchangeably in the following text.

EXAMPLE ONE: A PRIME PRODUCT WITH BYPRODUCTS

Situation

Seven operations have a network relationship.

One prime product and four byproducts are produced.

Three off grade products can be reworked to contribute to either the prime product or a byproduct.

Objective

Determine the material and labor costs for a SBQ of 1000 of the prime product. (The various calculations for this example are shown in Tables 5.1–5.9.)

TABLE 5.1   Quantity Per Required (for a SBQ of 1000)

Item	Quantity per
1	3000.00
2	1000.000
3	10.000
4	0.005

TABLE 5.2   Item Cost

Item	Cost ($) per
1	2.00
2	30.00
3	100.00
4	5000.00

TABLE 5.3   Labor Time (per Time Unit) Required (for a SBQ of 1000)

Operation	Time
10	10.0
10R	1.5
20	30.0
20R	10.0
30	20.0
30R	25.0
40	5.0

Labor rate is $15 per time unit

TABLE 5.4   Process Network

Operation	Input	Output	Yield (%)
10	1	10X	80
	2	10Y	20
10R	10Y	10X	60
		B1	40
20	3	20X	70
	10X	20Y	30
20R	20Y	20X	70
		B2	30
30	4	30X	60
	20X	30Y	40
30R	30Y	30X	80
		B3	20
40	30X	P	90
		B4	10

The P is the prime product.
The four Bs are the byproducts.
The Ys are products that go into rework.

TABLE 5.5   The Cumulative Yield Calculation

Operation	Input Yields			Output Product	Output Yield	Cumulative Yield
	1	2	Total			
10			1.00	10X	0.80	0.80
				10Y	0.20	0.20
10R	0.20		0.20	10X	0.60	0.12
				B1	0.40	0.08
20	0.80	0.12	0.92	20X	0.70	0.644
				20Y	0.30	0.276
20R	0.276		0.276	20X	0.70	0.1932
				B2	0.30	0.0828
30	0.644	0.1932	0.8372	30X	0.60	0.5023
				30Y	0.40	0.3349
30R	0.3349		0.3349	30X	0.80	0.2679
				B3	0.20	0.0670
40	0.5023	0.2679	0.7702	P	0.90	0.6932
				B4	0.10	0.0770

The process cumulative yield is 0.6932.

TABLE 5.6 The Material Cost Calculation

Operation	Input Item	Quantity per SBQ		Operation Input Yield		Process Cumulative Yield		Adjusted Quantity per	Item Cost ($) per	Total Item ($) Cost
10	1	3000.000	×	1.0000	=	3000.0000 / 0.6932	=	4327.7553	2.00	8,655.51
10	2	1000.000	×	1.0000	=	1000.0000 / 0.6932	=	1442.5851	30.00	43,277.55
20	3	10.000	×	0.9200	=	9.2000 / 0.6932	=	13.2718	100.00	1,327.18
30	4	0.005	×	0.8372	=	0.0042 / 0.6932	=	0.0061	5000.00	30.50
Total										$53,290.74

TABLE 5.7   The Labor Cost Calculation

Operation	Time per SBQ		Operation Input Yield		Process Cumulative Yield		Adjusted Time per
10	10.0	×	1.0000	=	10.0000 / 0.6932	=	14.4259
10R	1.5	×	0.2000	=	0.3000 / 0.6932	=	0.4328
20	30.0	×	0.9200	=	27.6000 / 0.6932	=	39.8153
20R	10.0	×	0.2760	=	2.7600 / 0.6932	=	3.9815
30	20.0	×	0.8372	=	16.7440 / 0.6932	=	24.1546
30R	25.0	×	0.3349	=	8.3725 / 0.6932	=	12.0780
40	5.0	×	0.7702	=	3.8510 / 0.6932	=	5.5554
Total time							100.4435
Labor rate per time unit						×	15.00
Total labor cost							$ 1506.65

TABLE 5.8   Total Labor and Material Costs

Material costs	=	$53,290.74
Labor costs	=	1,506.65
Total		$54,797.39

TABLE 5.9   Quantities of Products Produced

Product	Yield				Production Quantity
P	0.6932	×	1442.5851	=	1000.0000
B1	0.0800	×	1442.5851	=	115.4068
B2	0.0828	×	1442.5851	=	119.4460
B3	0.0670	×	1442.5851	=	96.6532
B4	0.0770	×	1442.5851	=	111.0791
Totals	1.0000				1442.5851

EXAMPLE TWO: A REWORK FEEDBACK LOOP WITH COPRODUCT

Situation

Five operations have a network relationship.

A rework operation causes a feedback loop.

Operation 10 produces As, Bs, and ABs.

Operation 20 first consumes As and Bs until they are depleted and then consumes ABs which can be used as either an A or a B.

An A (or an AB) plus a B (or an AB) with two 3s is required to make an X1.

Objective

Determine the adjusted quantities per for each operation. (The calculations for this example are shown in Tables 5.10 to 5.14.)

The Continuous Operation Mode

The key consideration for someone in this environment is to decide when to consume the reworked, in-stock As and Bs. The decision will probably be based on whether the SBQ cycle time is:

Short, as per hour or per shift

Long, as per week or per month

The following decision options will probably exist.

Consume in-stock A and B inventories in the next cycle and recalculate the required number of 1s and 2s.

Consume in-stock A and B inventories when they are sufficient to produce the required SBQ of X1s (2083.3332) without the need for operation 10.

TABLE 5.10 Quantity per Required (for a SBQ of 1000)

Item	Quantity per
1	2000
2	3000
3	2000
4	1000

TABLE 5.11   Process Network

Operation	Input	Output	Yield (%)
10 Chip making	1	A	20
	2	B	10
		AB	60
		S1	10
20 Initial assembly	3	X1	85
	A	S2	15
	B		
	AB		
	A1		
	B1		
30 Final assembly	4	X2	60
	X1	S3	10
		R	30
35 Rework	R	A1	40
		B1	60
XX Line stock for operation 20	A1	A1	100
	B1	B1	100
40 Final test	X2	P	80
		S4	20

P is the prime product.

The four Ss are scrap.

Final assembly produces Rs which are reworked, put into line stock and then used in operation 20.

TABLE 5.12   The Start Up Cumulative Yield Calculations

	Input			Output		
Operation	Product	Single Product Yield	Coproduct Yield	Product	Yield	Cumulative Yield
10		1.00		A	0.20	0.20
				B	0.10	0.10
				AB	0.60	0.60
				S1	0.10	0.10
20	A	0.20				
	B	0.10				
	AB	0.60				
	Total	0.90	0.45	X1	0.85	0.3825
				S2	0.15	0.0675
30	X1	0.3825		X2	0.60	0.2295
				R	0.30	0.1148
				S3	0.10	0.0383
35	R	0.1148		A1	0.40	0.0459
				B1	0.60	0.0689
40	X2	0.2295		P	0.80	0.1836
				S4	0.20	0.0459

The process cumulative yield is 0.1836

TABLE 5.13   The Start Up Adjusted Quantity per Calculation

Operation	Input Item	Quantity per SBQ	Operation Input Yield	Process Cumulative Yield	Adjusted Quantity per
10	1	2,000 ×	1.00 = 2,000	/ 0.1836	= 10,893.246
	2	3,000 ×	1.00 = 3,000	/ 0.1836	= 16,339.869
20	3	2,000 ×	0.45 = 900	/ 0.1836	= 4,901.961
30	4	1,000 ×	0.3825 = 382.5	/ 0.1836	= 2,083.333

The base quantity is calculated: 10,893.246/2 = 5,446.623.

TABLE 5.14   The Start Up Production Quantity Calculation

Product	Yield	Times Base Quantity Equals	Resultant In-Stock Quantity
S1	0.10	544.6623	
S2	0.0675	367.6471	
S3	0.0383	208.6057	
S4	0.0459	250.0000	
A1	0.0459	250.0000	250.0000
B1	0.0689	375.2723	375.2723
P	0.1836	1000.0000	
A	0.20	1089.3246	
B	0.10	544.6623	
AB	0.60	3267.9738	
X1	0.3825	2083.3332	

EXAMPLE THREE: REBUILD

Situation

Used items are returned to this company to be rebuilt.

When the input used item is disassembled, some percentage of the components are usable in the reworked product, while others must be obtained from new stock.

Objective

Determine the required production quantities.

TABLE 5.15   Quantity per Required (for a SBQ of 100)

Item	Quantity per	New Item (%)	Quantity
1	200	0.00	0.00
2	300	0.50	150
3	200	0.90	180
4	400	1.00	400
R1	100	1.00	100

TABLE 5.16   Process Network

Operation	Input	Output	Yield (%)
Source of rebuild stock		R1	100
10 Initial inspection	R1	R2	95
	R3	RS	5
20 Take apart	R2	Kit with:	
		1R	100
		2R	50
		3R	10
		2S	50
		3S	90
		4S	100
XX Line stock		2N	
		3N	
		4N	
30 Assemble	Kit:	X	100
	1R		
	2R		
	3R		
	2N		
	3N		
	4N		
40 Test	X	XT	80
		R3	20

R items (as 2R) are reusable.
S items (as 3S) are scrap.
N items (as 4N) are new.
RS items are scrap due to handling loss.

TABLE 5.17   The Cumulative Yield Calculations

	Input		Output		Cumulative
Operation	Product	Yield	Product	Yield	Yield
10	R1	1.00	R2	0.95	0.95
			RS	0.05	0.05
20	R2	0.95	Kit:	1.00	0.95
			2S	0.50	0.475
			3S	0.90	0.855
			4S	1.00	0.95
30	KIT	0.95	X	1.00	0.95
40	X	0.95	XT	0.80	0.76
			R3	0.20	0.19

The process cumulative yield is 0.76

TABLE 5.18   The Adjusted Quantity per Calculation

Operation	Input Item	Quantity per SBQ		Operation Input Yield			Process Cumulative Yield		Adjusted Quantity per
10	R1	100	×	1.00	=	100.0	/ 0.76	=	132
30	2N	150	×	0.95	=	142.5	/ 0.76	=	188
	3N	180	×	0.95	=	171.0	/ 0.76	=	225
	4N	400	×	0.95	=	380.0	/ 0.76	=	500

The base quantity if equal to R1 or 132.

TABLE 5.19   The Production Quantity Calculation

Product	Yield	Times Base Quantity Equals
RS	0.05	6.6
2S	0.475	62.7
3S	0.855	112.86
4S	0.95	125.4
XT	0.76	100.32
R3	0.19	25.08

EXAMPLE FOUR: REWORK

Situation

Product D is produced on a continuous flow, repetitive line.

After test, some of the product can be disassembled so that part A can be salvaged and used again on the line.

The line is balanced.

Objective

Identify the quantity of components required each hour on the line.

Rates

The Daily Going Rate (DGR) is 800.

The hourly rate is 100.

(The calculations for this example are shown in Tables 5.20 and 5.21.)

Process Network Analysis

Due to dual A inputs at operation 20, the illustrated process must be viewed as two separate processes.

Yields and adjusted quantities must be calculated for operations 20 through 45 (process A) to identify the quantity of As that are required from operation 10. Operation 10 quantities are then calculated (process B).

(The calculations for the process network analysis for this example are shown in Tables 5.22 to 5.27.)

TABLE 5.20   Component Requirements

The component quantities required to make Ds are:

Item	Quantity per D	Quantity per Hour
1	3	[a]
2	2	[a]
3	1	100
4	1	100
5	1	100

[a]These items will be calculated later.

TABLE 5.21   The Process Network

Operation	Input	Output	Yield (%)
10 Assemble	1	A	100
	2		
20 Assemble	3	B	100
	A		
	A1		
30 Assemble	4	C	90
	B	CS	10
40 Test	5	D	80
	C	DR	20
45 Take apart	DR	A1	100
		3S	100
		4S	100
		5S	100
XX WIP Stores	A1	A1	100

The S items are scrap.
A1s and As are actually the same physical items.

TABLE 5.22   The Process A Cumulative Yield Calculation

Operation	Input		Output		Cumulative Yield
	Product	Yield	Product	Yield	Yield
20		1.00	B	1.00	1.00
30	B	1.00	C	0.90	0.90
			CS	0.10	0.10
40	C	0.90	D	0.80	0.72
			DR	0.20	0.18
45	DR	0.18	A1	1.00	0.18
			3S	1.00	0.18
			4S	1.00	0.18
			5S	1.00	0.18

The process A cumulative yield is 0.72.

TABLE 5.23   The Process A Adjusted Quantity Per Calculation

Operation	Input Item	Hourly DGR Quantity			Operation Input Yield			Process Cumulative Yield			Adjusted Quantity per
20	3	100	×	1.00	=	100	/ 0.72	=	139
	A	100	×	1.00	=	100	/ 0.72	=	139
30	4	100	×	1.00	=	100	/ 0.72	=	139
40	5	100	×	0.90	=	90	/ 0.72	=	125

The process A base quantity is 139.0.

TABLE 5.24   The Process A Production Quantity Calculation

Product	Yield	Times Base Quantity Equals
CS	0.10	14
3S	0.18	25
4S	0.18	25
5S	0.18	25
D	0.72	100
A1	0.18	25

The results of process A:
   The production of 100 Ds also produces 25 A1s, which means that the number of A1s required from process B is 139 − 25 = 114 (hourly DGR quantity). The hourly rate for process B is 114.
The component quantities to make As are:

Item	Quantity per A	Quantity per Hour
1	3	342
2	2	228

TABLE 5.25   The Process A Cumulative Yield Calculation

Operation	Input Product	Input Yield	Output Product	Output Yield	Cumulative Yield
10		1.00	A	1.00	1.00

The process A cumulative yield is 1.00

TABLE 5.26 The Process B Adjusted Quantity per Calculation

Operation	Input Item	Hourly DGR Quantity		Operation Input Yield			Process Cumulative Yield			Adjusted Quantity per
10	1	342	×	1.00	=	342	/ 1.00	=		342
	2	228	×	1.00	=	228	/ 1.00	=		228

TABLE 5.27 The Summary of Hourly Required Components

Item	Required Quantity
1	342
2	228
3	139
4	139
5	125

The previous four examples were provided to illustrate the total capability that is required for yield management.

When yield variations occur at the work station, milestone, or line level, yield calculations are necessary to determine the material "adjusted quantities per" that are required to produce product. This is primarily an interface to material requirements planning. The previous four examples illustrate the types of calculations that are required.

As product designs and production techniques are refined, variations with yields will occur. Although these variations are often thought of as short lived, in actuality they may exist for weeks or months. It is essential to understand in the planning systems that these yield trends exist and to apply them to the planned production quantities. Yield trends must be calculated and applied to production planning, master schedule planning, material requirements planning, and setup planning.

CHAPTER SIX

LINE BALANCING

THE GENERAL PURPOSE OF LINE BALANCING

The objective of line balancing is to minimize the cost of operating a flow line by optimizing the total line-idle time cost to a minimum. It is an activity normally performed by industrial and/or manufacturing engineering.

The line balance process requires the following inputs:

The DGR or daily schedule to produce a product or a family of products.

The task relational network for the products/models/options with associated dependencies and task times. Either a single network of all products or multiple networks (one per product) may be used to depict task dependencies.

Available labor to perform the required tasks.

Machine constraints such as: a turntable, with a press for option A and a press for option B, is bolted to the floor at work station four, or an air screwdriver needs to be plugged into station six.

Work content for a line, where the tasks have been preassigned to work stations, and the objective is to assign unassigned tasks and adjust the mix quantities of the models on the line.

Line balancing applies to transfer lines, flow lines, repetitive lines, assembly lines, JIT lines, and continuous flow manufacturing (of discrete items) lines.

Line balancing is performed when:

The DGR changes.

The product design is initiated/changed.

The mix of products/models/options within the family to be balanced on the line changes.

The available labor skills (quantity) change.

The available machines and/or tools change.

DEFINITION OF TERMS

Assembly Line. A series of work stations on which one or more tasks are performed by people and/or machines to produce a product or products.

Balance Delay. An arithmetic calculation used to determine a balanced line. The smallest balance delay is considered to be optimum. The calculation is as follows:

$$\text{Balance Delay} = \frac{\text{Total Idle Time}}{(\text{Number of Stations}) \times (\text{Cycle Time})} \times 100$$

Balanced Line. A line with a minimum idle time (a quantity) or a minimum balance delay (a percentage).

Cycle Time. The amount of time that a product spends at a work station. The cycle time for any one work station is the same for all work stations.

Flow Line. See "Assembly Line."

Idle Time. The amount of cycle time at a work station during which no tasks (work) are being performed.

Repetitive Line. See "Assembly Line."

Task. A measurable element of work required to produce a product.

Task Factor. A calculation used to determine which task should be assigned next to a work station. It is equal to: one, plus the number of immediate successors, times the task time. When multiple tasks are available for assignment to a work station, the task with the largest task factor is assigned first.

Task Network. A dependency relationship showing which tasks must be preceded by which other tasks. It is frequently illustrated by a drawing such as a critical path network.

Task Time. The time necessary to perform a task. All tasks in the same task network must use the same unit of measure for time (i.e., seconds, minutes, or hours) that is used for the cycle time.

Work Station. A position on a flow line where one or more tasks may be performed.

SOME GENERAL PARAMETERS

The time used throughout the process is often formatted as XXX.XXX, and means seconds, minutes, or hours based on the user's discretion.

A single task network may contain a large number of tasks, such as up to 999.

A task network may have multiple starting nodes which may be feeder lines.

A task network may have multiple ending nodes if multiple products are being produced on the line and all task definitions have been combined into a single task network.

Every task in a task network must have a unique task number. If the same work (task) is performed at multiple places in the task network, each occurrence of the work must be identified by a unique task number.

The tasks within a task network may be numbered in any order and do not have to relate to the sequence of the network.

INTRODUCTION TO A LINE BALANCING EXAMPLE

Assume that seven tasks need to be performed to produce a product. The seven tasks do not necessarily have to be performed in a sequential (1–7) manner. The task relationships are shown in Table 6.1.

The network relationship of tasks can be illustrated by Figure 6.1.

This network shows that task 3 can be performed any time between task 1 and task 5. Task 6 can be started when both tasks 4 and 5 are completed and before task 7 starts.

Figure 6.1 illustrates a slight logic flaw in that it is redundant on some relationships. Task 5 requires both tasks 2 and 4 to be completed before it starts, yet task 4 cannot start until task 2 has completed. The dependency from task 2

TABLE 6.1 Task Relationships

Task Number	Task Time	Preceded by
1	8.5	None
2	4.0	1
3	3.2	1
4	2.0	2
5	5.6	2,3,4
6	1.8	4,5
7	2.5	6

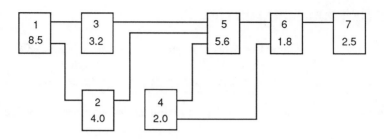

FIGURE 6.1   The Task Network

to task 5 is redundant. Another redundancy exists between tasks 4, 5, and 6. The construction of a task network with built-in redundancy is a common thing, since many times the redundancy is not always obvious. However, since some people might argue that these additional redundant dependencies will alter the results, we will perform the example in two passes, with and without the redundancy, and show that the results are identical.

This first pass will be performed with the redundant dependencies.

The seven tasks could be performed on some number of work stations ranging from one to seven. The cycle time will fall within these ranges.

> Minimum cycle time will be equal to the largest task time, or 8.5. Note that this would not be true if a mix of products were being built on the line, since the interspersed quantity of each product would have to be considered.
>
> Maximum cycle time will be equal to the sum of all of the task times (8.5 + 4.0 + 3.2 + 2.0 + 5.6 + 1.8 + 2.5), or 27.6.

The system may, in this example, be constrained by the specification of:

> The starting cycle time.
> The minimum number of work stations to be considered.
> The cycle time iteration value (by how much should the cycle time be increased for the next iteration).

INITIAL DATA ENTRY

To initiate processing for the example, a user would input the following data.

> *Starting Cycle Time.* The starting cycle time is set to 8.5.
>
> *Minimum Number of Work Stations.* The minimum number of work stations is set to 2.
>
> *Cycle Time Iteration Value.* The cycle time is set to increase by a value of 2.0 for each iteration.

TABLE 6.2   Task Data

Number	Name	Task Time	Preceded by	Followed by
1	A	8.5	None	2,3
2	B	4.0	1	4,5
3	C	3.2	1	5
4	D	2.0	2	5,6
5	E	5.6	2,3,4	6
6	F	1.8	4,5	7
7	G	2.5	6	None

INITIAL SYSTEM STEPS

The system will perform the initial steps shown in Table 6.3 to 6.5.

TABLE 6.3   Calculate Task Factors

Task Number	Number of Task Successors	Plus One Equals	Times Task Time	Equals Task Factor
1	2	3	8.5	25.5
2	2	3	4.0	12.0
3	1	2	3.2	6.4
4	2	3	2.0	6.0
5	1	2	5.6	11.2
6	1	2	1.8	3.6
7	0	1	2.5	2.5

TABLE 6.4   Construct a Task Table Sequenced by the Task Factors

Task Table

Task Number	Task Factor	Task Time	Predecessor Tasks 1	2	3	4	5	6	7	Available
1	25.5	8.5								X
2	12.0	4.0	X							
5	11.2	5.6		X	X	X				
3	6.4	3.2	X							
4	6.0	2.0		X						
6	3.6	1.8				X	X			
7	2.5	2.5						X		

Note: Tasks that do not have any predecessor tasks are classed as being available.

TABLE 6.5   Construct a Work Station Table

Work Station Table

Work	Assigned Tasks							Consumed	Unconsumed
Station	1	2	3	4	5	6	7	Time	Time (Idle)
1								0.0	8.5

ITERATION CALCULATIONS

In the task table, find the task with the largest task factor. Only task 1 is available.

Subtract the task time from the unconsumed time in the work station table. If the answer is positive, add the task to the work station table and delete it from the task table, as shown in Table 6.6.

Test to see if tasks are available. If they are, find the available task with the largest task factor. Tasks 2 and 3 are now available since task 1 has been assigned to the work station table and deleted from the task table. Task 2 has the largest task factor.

Subtract the task time of task 2 from the unconsumed time of station 1. If the answer is negative, test for a fit with the next available task (3) that has the

TABLE 6.6   Step One

Work Station Table

Work	Assigned Tasks							Consumed	Unconsumed
Station	1	2	3	4	5	6	7	Time	Time (Idle)
1	8.5							8.5	0.0

Task Table

Task	Task	Task	Predecessor Tasks							
Number	Factor	Time	X	2	3	4	5	6	7	Available
2	12.0	4.0								X
5	11.2	5.6		X	X	X				
3	6.4	3.2								X
4	6.0	2.0		X						
6	3.6	1.8				X	X			
7	2.5	2.5						X		

largest factor. Neither task 2 or 3 will fit into station 1 since the unconsumed time if 0.0.

When all available tasks have been tested for a fit into a work station and none fit, the work station number is incremented, and the process is repeated. This provides the results shown in Table 6.7 for station 2.

TABLE 6.7 Step Two

Work Station Table

Work Station	Assigned Tasks							Consumed Time	Unconsumed Time (Idle)
	1	2	3	4	5	6	7		
1	8.5							8.5	0.0
2		4.0						4.0	4.5

Task Table

Task Number	Task Factor	Task Time	Predecessor Tasks							Available
			X	X	3	4	5	6	7	
5	11.2	5.6			X	X				
3	6.4	3.2								X
4	6.0	2.0								X
6	3.6	1.8				X	X			
7	2.5	2.5						X		

TABLE 6.8 Step Three

Work Station Table

Work Station	Assigned Tasks							Consumed Time	Unconsumed Time (Idle)
	1	2	3	4	5	6	7		
1	8.5							8.5	0.0
2		4.0	3.2					7.2	1.3

Task Table

Task Number	Task Factor	Task Time	Predecessor Tasks						Available
			X	X	4	5	6	7	
5	11.2	5.6			X				
4	6.0	2.0							X
6	3.6	1.8			X	X			
7	2.5	2.5					X		

Test to see if tasks are available. If they are, find the task with the largest task factor. Tasks 3 and 4 are now available, and task 3 has the largest task factor.

Subtract the task time from the unconsumed time for task 3 at station 2. Task 3 with a time of 3.2 will fit within the unconsumed time of 4.5. The updated tables appear as shown in Table 6.8.

Test to see if tasks are available. Task 4 is now available.

Attempt to subtract the task time of task 4 from the unconsumed station time. Task 4 will not fit into station 2, so the station number must be incremented. Task 4 is assigned as shown in Table 6.9.

The next fit test, shown in Table 6.10, assigns task 5 to station 3.

To assign the available task 6 requires that the work station be incremented to number 4. Both task 6 and 7 will fit into station 4 as shown in Table 6.11.

At this point, no tasks are available for assignment and the balance delay needs to be calculated. The arithmetic equations are shown below.

$$\text{Balance Delay} = \frac{\text{Idle Time}}{(\text{Number of Stations})(\text{Cycle Time})} \times 100$$

$$\text{Balance Delay} = \frac{0.0 + 1.3 + 0.9 + 4.2}{4 \times 8.5} \times 100$$

$$\text{Balance Delay} = 18.8\%$$

The first iteration with the known redundancies is now complete.

TABLE 6.9 Step Four

Work Station Table

Work Station	\multicolumn{7}{c}{Assigned Tasks}	Consumed Time	Unconsumed Time (Idle)
	1	2	3	4	5	6	7		
1	8.5							8.5	0.0
2		4.0	3.2					7.2	1.3
3				2.0				2.0	6.5

Task Table

Task Number	Task Factor	Task Time	\multicolumn{7}{c}{Predecessor Tasks}	Available
			X	X	X	X	5	6	7	
5	11.2	5.6								X
6	3.6	1.8					X			
7	2.5	2.5						X		

TABLE 6.10   Step Five

Work Station Table

| Work | Assigned Tasks | | | | | | | Consumed | Unconsumed |
Station	1	2	3	4	5	6	7	Time	Time (Idle)
1	8.5							8.5	0.0
2		4.0	3.2					7.2	1.3
3				2.0	5.6			7.6	0.9

Task Table

| Task | Task | Task | Predecessor Tasks | | | | | | | |
Number	Factor	Time	X	X	X	X	X	6	7	Available
6	3.6	1.8								X
7	2.5	2.5						X		

TABLE 6.11   Step Six

Work Station Table

| Work | Assigned Tasks | | | | | | | Consumed | Unconsumed |
Station	1	2	3	4	5	6	7	Time	Time (Idle)
1	8.5							8.5	0.0
2		4.0	3.2					7.2	1.3
3				2.0	5.6			7.6	0.9
4						1.8	2.5	4.3	4.2

Task Table

| Task | Task | Predecessor Tasks | | | | | | | |
Number	Factor	X	X	X	X	X	X	X	Available

This next pass will be performed without the redundant dependencies. Figure 6.2 illustrates the revised network.

Since the previous logic is identical to what will now be employed, only the step-by-step Tables 6.12–6.22 will be shown.

Note that although variations existed in the beginning corrected tables (compared to the ones with the redundant dependencies), the latter tables and the final results are identical.

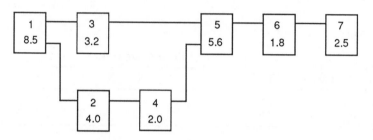

FIGURE 6.2   Corrected—The Task Network

TABLE 6.12   Corrected – Task Relationships

Task Number	Task Time	Preceded by
1	8.5	None
2	4.0	1
3	3.2	1
4	2.0	2
5	5.6	3,4
6	1.8	5
7	2.5	6

TABLE 6.13   Corrected – Task Data

Number	Name	Task Time	Preceded by	Followed by
1	A	8.5	None	2,3
2	B	4.0	1	4
3	C	3.2	1	5
4	D	2.0	2	5
5	E	5.6	3,4	6
6	F	1.8	5	7
7	G	2.5	6	None

TABLE 6.14 Corrected—Calculate Task Factors

Task Number	Number of Task Successors	Plus One Equals	Times Task Time	Equals Task Factor
1	2	3	8.5	25.5
2	1	2	4.0	8.0
3	1	2	3.2	6.4
4	1	2	2.0	4.0
5	1	2	5.6	11.2
6	1	2	1.8	3.6
7	0	1	2.5	2.5

TABLE 6.15 Corrected—Construct a Task Table Sequenced by the Task Factors

Task Table

Task Number	Task Factor	Task Time	Predecessor Tasks 1	2	3	4	5	6	7	Available
1	25.5	8.5								X
5	11.2	5.6		X	X					
2	8.0	4.0	X							
3	6.4	3.2	X							
4	4.0	2.0		X						
6	3.6	1.8				X				
7	2.5	2.5					X			

Note: Tasks that do not have any predecessor tasks are classed as being available.

TABLE 6.16 Corrected—Construct a Work Station Table

Work Station Table

Work Station	Assigned Tasks 1	2	3	4	5	6	7	Consumed Time	Unconsumed Time (Idle)
1								0.0	8.5

TABLE 6.17 Corrected — Step One

Work Station Table

Work Station	Assigned Tasks							Consumed Time	Unconsumed Time (Idle)
	1	2	3	4	5	6	7		
1	8.5							8.5	0.0

Task Table

Task Number	Task Factor	Task Time	Predecessor Tasks							Available
			X	2	3	4	5	6	7	
5	11.2	5.6			X	X				
2	8.0	4.0								X
3	6.4	3.2								X
4	4.0	2.0	X							
6	3.6	1.8					X			
7	2.5	2.5						X		

TABLE 6.18 Corrected — Step Two

Work Station Table

Work Station	Assigned Tasks							Consumed Time	Unconsumed Time (Idle)
	1	2	3	4	5	6	7		
1	8.5							8.5	0.0
2		4.0						4.0	4.5

Task Table

Task Number	Task Factor	Task Time	Predecessor Tasks							Available
			X	X	3	4	5	6	7	
5	11.2	5.6			X	X				
3	6.4	3.2								X
4	4.0	2.0								X
6	3.6	1.8					X			
7	2.5	2.5						X		

TABLE 6.19   Corrected—Step Three

Work Station Table

Work Station	Assigned Tasks							Consumed Time	Unconsumed Time (Idle)
	1	2	3	4	5	6	7		
1	8.5							8.5	0.0
2		4.0	3.2					7.2	1.3

Task Table

Task Number	Task Factor	Task Time	Predecessor Tasks						Available
			X	X	4	5	6	7	
5	11.2	5.6			X				
4	4.0	2.0							X
6	3.6	1.8				X			
7	2.5	2.5					X		

TABLE 6.20   Corrected—Step Four

Work Station Table

Work Station	Assigned Tasks							Consumed Time	Unconsumed Time (Idle)
	1	2	3	4	5	6	7		
1	8.5							8.5	0.0
2		4.0	3.2					7.2	1.3
3				2.0				2.0	6.5

Task Table

Task Number	Task Factor	Task Time	Predecessor Tasks						Available
			X	X	X	X	5	6	7	
5	11.2	5.6								X
6	3.6	1.8					X			
7	2.5	2.5						X		

TABLE 6.21   Corrected — Step Five

Work Station Table

Work Station	Assigned Tasks							Consumed Time	Unconsumed Time (Idle)
	1	2	3	4	5	6	7		
1	8.5							8.5	0.0
2		4.0	3.2					7.2	1.3
3				2.0	5.6			7.6	0.9

Task Table

Task Number	Task Factor	Task Time	Predecessor Tasks							Available
			X	X	X	X	X	6	7	
6	3.6	1.8								X
7	2.5	2.5						X		

TABLE 6.22   Corrected — Step Six

Work Station Table

Work Station	Assigned Tasks							Consumed Time	Unconsumed Time (Idle)
	1	2	3	4	5	6	7		
1	8.5							8.5	0.0
2		4.0	3.2					7.2	1.3
3				2.0	5.6			7.6	0.9
4						1.8	2.5	4.3	4.2

Task Table

Task Number	Task Factor	Predecessor Tasks							Available
		X	X	X	X	X	X	X	

TABLE 6.23 Summary of Remaining Iterations

Iteration	Cycle Time	Work Station	Assigned Tasks							Idle Time	Balance Delay (%)
			1	2	3	4	5	6	7		
1	8.5	1	8.5							0.0	
		2		4.0	3.2					1.3	
		3				2.0	5.6			0.9	
		4						1.8	2.5	4.2	
										6.4	18.8
2	10.5	1	8.5							2.0	
		2		4.0	3.2	2.0				1.3	
		3					5.6	1.8	2.5	0.6	
										3.9	12.4
3	12.5	1	8.5	4.0						0.0	
		2			3.2	2.0	5.6			1.7	
		3						1.8	2.5	8.2	
										9.9	26.4
4	14.5	1	8.5	4.0		2.0				0.0	
		2			3.2		5.6	1.8	2.5	1.4	
										1.4	4.8
5	16.5	1	8.5	4.0	3.2					0.8	
		2				2.0	5.6	1.8	2.5	4.6	
										5.4	16.4

TABLE 6.23 *Continued*

Iteration	Cycle Time	Work Station	1	2	3	4	5	6	7	Idle Time	Balance Delay (%)
			\multicolumn{7}{}{Assigned Tasks}								
6	18.5	1	8.5	4.0	3.2	2.0				0.8	
		2					5.6	1.8	2.5	8.6	
										9.4	25.4
7	20.5	1	8.5	4.0	3.2	2.0				2.8	
		2					5.6	1.8	2.5	10.6	
										13.4	32.7
8	22.5	1	8.5	4.0	3.2	2.0				4.8	
		2					5.6	1.8	2.5	12.6	
										17.4	38.7
9	24.5	1	8.5	4.0	3.2	2.0	5.6			1.2	
		2						1.8	2.5	20.2	
										21.4	43.7
10	26.5	1	8.5	4.0	3.2	2.0	5.6	1.8		1.4	
		2							2.5	24.0	
										25.4	47.9
11	28.5	1	8.5	4.0	3.2	2.0	5.6	1.8	2.5	0.9	
										0.9	3.2

To begin the next iteration, the cycle time incremental value of 2.0 is added to the original 8.5 cycle time. However, since the logic for all iterations is identical, only the summarized data of the iterations is shown in Table 6.23.

Note that iteration 11 is invalid since it exceeded the specified minimum of two work stations. The smallest balance delay is found in iteration 4 at 4.8%.

This has been a simple example of line balancing which was provided so that the reader could gain an initial understanding of the balancing process. Further details and logic variations may be found in the *Industrial Engineering Handbook* by G. Salvendy, Wiley, 1982.

Line balancing is required for the labor intensive environment where the line is basically a belt, variations in available labor are common, and the work stations can be set up for task performance in a few hours or less. The previous example applies to this environment.

The function applies to a flow line with physical constraints, facilities with flexible manufacturing cells, and those with a stable labor force. Although it does not apply to a pure job shop, it does apply to that job shop which is moving toward the lights-out factory concept and has applied flexible manufacturing techniques.

PART THREE

BUILDING A MIXED JOB-SHOP/FLOW-SHOP PRODUCTION EXAMPLE

CHAPTER SEVEN

ESTABLISHING CRITERIA FOR THE CASE STUDY EXAMPLE

INTRODUCING THE CASE STUDY EXAMPLE

The simple example that is defined in this section, and used in following sections, has been created to illustrate the top level planning logic for production and master scheduling systems in a mixed job-shop/flow-shop manufacturing environment.

Every effort has been made to constrain the example with a basic rule—Keep it simple so that everyone can understand it. For clarification, some basic assumptions are defined, and some terms have been restated so that their meaning is clearly understood as related to the example.

In addition to the assumptions, some conditions are defined that have been applied to the example for simplification purposes. They were developed for this example, and may not necessarily be available in a specific software product which might have been developed to address this manufacturing environment.

SPECIFIC ASSUMPTIONS FOR THE CASE STUDY EXAMPLE

Some production facilities are illustrated on the following pages. They are not necessarily in balance or optimum for the production of the required end-item quantities. The assumption is that most companies cannot afford the luxury of a perfectly balanced plant.

The flow line production rates can deviate (for minimums and maximums) by plus or minus 10 % from the desired rates. Requirements that may occur for more than the maximum rates will not allow for demands to be met. Requirements for less than the minimum rates may mean that the line will be shut down.

Although it is common for demands to change during planning periods (or even daily and hourly), this example considers demands to be stable for the planning cycle while production plans are developed by planning periods.

Many purchased components are illustrated in the example. These would normally be of critical concern (due to costs), but the planned quantities are not developed in the example since the objective is to prepare manufacturing production plans and schedules.

In this example (and in most real-life situations), an objective is to have the production plans as level as possible. This basic rule will have variations depending on company policies regarding the full employment of labor and the degree to which building excess inventory will be permitted.

DEFINITION OF TERMS USED IN THE CASE STUDY EXAMPLE

End Item. The top level of the product that requires planning is not necessarily what ships to the customer. For example, the item may be a key assembly, or service part (Field Replaceable Unit—FRU). The end item is usually considered to be the top level of the item that requires planning. In this specific example, all end items are the finished products that ship to the customer.

Flow Shop. A shop that schedules the production of discrete items on a continuous basis using item numbers and a quantity required per period, which is usually referred to as a rate, such as a daily going rate.

Hedge Inventory. In a seasonal demand environment, that inventory which is built earlier than needed because capacity during the peak demand periods is not expected to be available.

Job Shop. A shop that schedules the production of discrete items on an intermittent basis using shop order numbers and lot sized quantities.

Product Family. A group of end items that requires about the same amount of the same resources during production. Note that while this is the product family definition that is used in this example, there are many acceptable approaches to product family definitions beyond this basic one.

Production Schedule. The quantity of items to be produced over a specific time period to meet a specific demand.

Rate. The quantity of an item which can be produced within a specified time period as, 100 per day, or 500 per week. Three types of rates are normally considered.

Minimum Rate. The least amount of product that can be produced per time period for the process to be cost effective. This is often 1 to 15% less than the desired rate.

Desired Rate. The rate set by management at which they would like the facility to operate. It is usually based on the facility design specifications.

Maximum Rate. The most amount of product that can be produced per time period for the process due to a physical constraint, such as the belt will not go any faster. This is often 1 to 15% more than the desired rate.

CONDITIONS APPLIED TO SIMPLIFY THE CASE STUDY EXAMPLE

The following conditions have been applied to the example to better illustrate the planning logic without the added complexities created by many arithmetic calculations and tables.

The Required Quantity Per is always equal to one.

Queue, move, and setup times are not used to reduce the arithmetic calculations, even though some data of this type may be illustrated in some sections.

Each work center has only one work station or machine, except for work center two which has two work stations.

A work station may perform one or more operations or tasks on items passing through the work station.

The sequencing of how product families are to be scheduled on the flow lines (based on line changeover costs) is assumed to have been performed external to the example.

Where multiple items are produced on a single flow line, item sequencing (based on work station changeover costs) and optimum item run quantity calculations (such as: five As, followed by two Bs, followed by five As, and so on) are assumed to have been performed external to this example.

Although not illustrated as such, stores exists as a warehouse with receipts and issues, in which both subassemblies and purchased parts may be stocked. Items shown as existing in stores (in the example) are considered to be purchased parts. The in-and-out flow of subassemblies is not illustrated in the following example diagrams.

Items that are shown as moving from one facility to another may or may not pass in and out of stores during the indicated moves.

Stores does not necessarily surround the plant facilities as shown in the following diagrams. It is shown that way to simplify the illustrations.

All lot-sized items are assumed to have an on-hand available balance of 50% of the lot size at the start up of the example.

At the start up of the example, it is assumed that a previous production plan does not exist.

A specific product is always built on a specific line or in a specific work center in the example. In real life, a planning system would accommodate alternate facilities (work stations, flow lines, work centers, subcontractors) with different run times for scheduling alternatives.

The key planning horizon used in the example is a 10-period horizon used for hedge inventory accumulation and blended demand calculations. An actual planning system would have horizons such as:

A horizon for actual orders only

A horizon for a blend of actual orders and forecasted demand

A horizon for forecasted demand only

A horizon for hedge inventory accumulation

To simplify the example, no offset planning is used. This means that an order's start date and finish data are always the same date.

Lot-sized quantities are planned based on the item's specified lot size or a multiple of that lot-size quantity.

This chapter has defined some basic criteria for the example. Without these constraints to maintain simplicity, this book would have to be published as a set of a dozen volumes.

The next chapter defines the production environment of the example. The example's assumptions and environment must be clearly understood for the planning logic in the following sections to reflect a reasonable solution to the top level planning problem.

CHAPTER EIGHT

CHARACTERISTICS OF THE PRODUCTION EXAMPLE

FAMILY PRODUCT STRUCTURES

The end items have been structured into product families as shown in Figure 8.1. A product family structure consists of those items that consume about the same amount of the same resources.

Four families have been constructed from the nine end items.

END ITEM PRODUCT STRUCTURES

Each end item (by a family grouping) consists of the assembly and component structures shown in Figure 8.2.

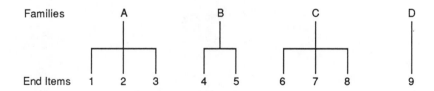

FIGURE 8.1   The Structure of End Items into Families

Family A

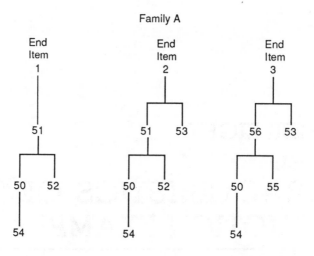

Family B

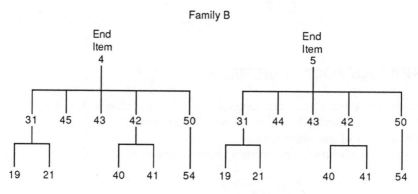

FIGURE 8.2 End Item Product Structures

Family C

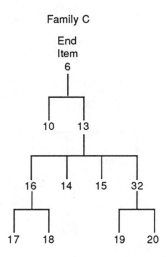

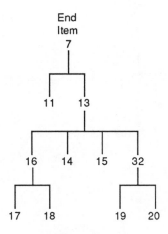

FIGURE 8.2 (continued)

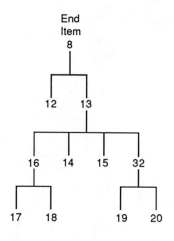

Family D

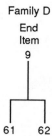

FIGURE 8.2 (continued)

PLANT LAYOUT OF THE FACILITIES

Figure 8.3 illustrates the example facility layout of four work centers, five flow lines, and the surrounding stores.

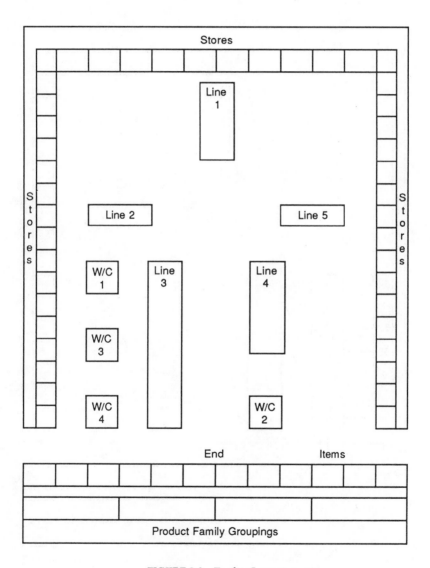

FIGURE 8.3   Facility Layout

MATERIAL FLOWS FOR PRODUCT FAMILIES

Figures 8.4–8.7 illustrate how material flows through the indicated facilities. It will start with the material flows of family A and then progressively add the additional families.

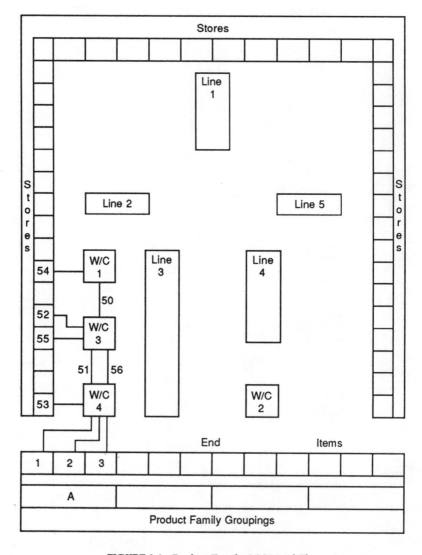

FIGURE 8.4   Product Family A Material Flow

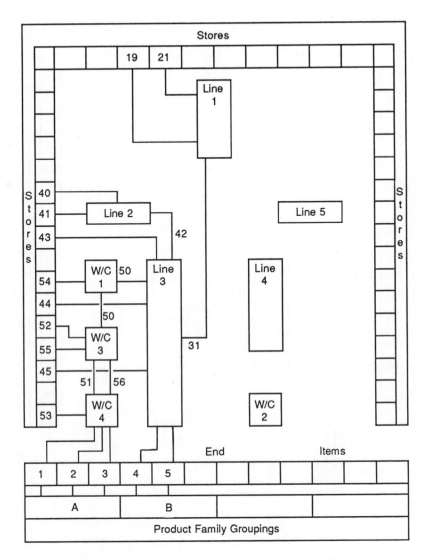

FIGURE 8.5 Product Family Material Flow B Added to Flow A

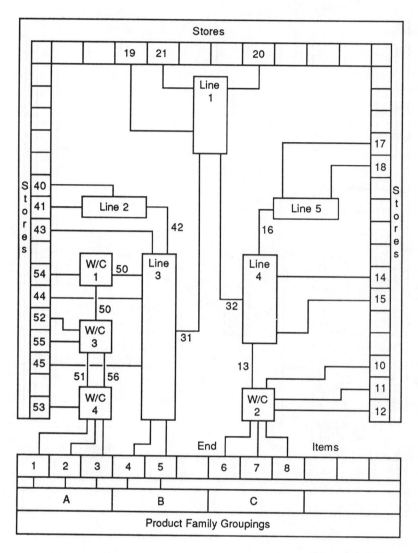

FIGURE 8.6 Product Family Material Flow C Added to Flows A and B

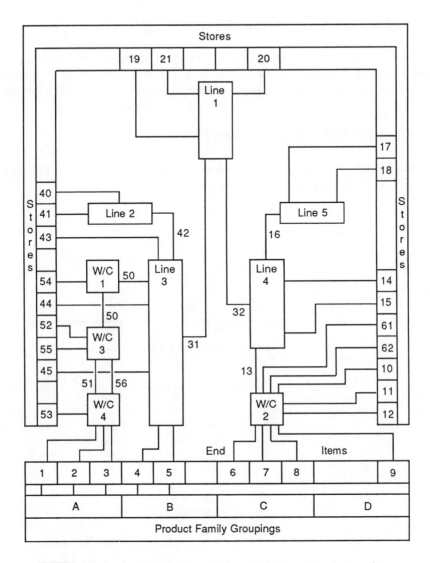

FIGURE 8.7 Product Family Material Flow D Added to Flows A, B, and C

ITEM AND FACILITY PLANNING STATISTICS

Table 8.1 shows some of the basic statistics that are used for planning purposes. Note that not all of the following data is used in the example. The additional data is available if the reader wishes to expand the case study example.

FLOW LINE OPERATIONAL LIMIT RATES

Table 8.2 describes the minimum and maximum limits that constrain the example flow line rates (at plus or minus 10%).

TABLE 8.1 Basic Planning Statistics for the Example

Facility	Daily Capacity (4)	Item	Setup Time (4)	Lot Size (5)	Lot Run Time (4)	Item Run Time (4)	Daily Run Quantity	Notes
W/C 1	480	50	0	5,000	480.	0.096	5,000	(2)
W/C 2	960	6	20	200	216.	1.08	444	
		7	30	300	324.	1.08	444	
		8	50	500	420.	0.84	571	(1)
		9	20	200	120.	0.6	800	
W/C 3	480	51	0	100	420.	4.2	114	
		56	0	100	420.	4.2	114	
W/C 4	480	1	1	10	60.	6.	80	
		2	2	20	120.	6.	80	
		3	3	30	180.	6.	80	
Line 1	480	31	0	1	0.32	0.32	1,500	(3)
		32	0	1	0.32	0.32	1,500	(3)
Line 2	480	42	0	1	0.48	0.48	1,000	(2)
Line 3	480	4	0	1	0.24	0.24	2,000	(3)
		5	0	1	0.96	0.96	500	(3)
Line 4	480	13	0	1	0.96	0.96	500	(2)
Line 5	480	16	0	1	0.048	0.048	10,000	(2)

Notes:
(1) Constrained by line 4 to a quantity of 500.
(2) Dedicated facility for a specific product.
(3) Interspersed products produced on the line.
(4) Time in minutes.
(5) A lot size may be the indicated amount or any multiple of that amount.

TABLE 8.2   Flow Line Rates

Rates per one day period

Line	Item	Minimum Rate	Desired Rate	Maximum Rate
1	31	1,350	1,500 (1)	1,650
	32	1,350	1,500 (1)	1,650
2	42	900	1,000	1,100
3	4	1,800	2,000 (2)	2,200
	5	450	500 (2)	550
4	13	450	500	550
5	16	9,000	10,000	11,000

Rates per five day period

Line	Item	Minimum Rate	Desired Rate	Maximum Rate
1	31	6,750	7,500 (3)	8,250
	32	6,750	7,500 (3)	8,250
2	42	4,500	5,000	5,500
3	4	9,000	10,000 (3)	11,000
	5	2,250	2,500 (3)	2,750
4	13	2,250	2,500	2,750
5	16	45,000	50,000	55,000

Notes:

(1) 1500 item 31s, or 1500 item 32s, or 1500 of a mix of items 31 and 32 is the desired daily rate for the line.

(2) 2000 item 4s, or 500 items 5s, or some quantity of a mixture of items 4 and 5 (the quantity being dependent on the mixture) is the desired daily rate for the line.

(3) Notes (1) and (2) should be considered to be extended for the five-day planning periods.

TABLE 8.3 Gross Demand

Item	Period									
	1	2	3	4	5	6	7	8	9	10
1	10	30	50	80	70	60	50	40	30	20
2	200	200	200	200	200	200	200	200	200	200
3	140	130	120	100	120	140	160	180	200	220
4	1000	1100	1200	1300	1400	1500	1600	1700	1800	1900
5	2000	1900	1800	1700	1600	1400	1300	1200	1100	1000
6	500	600	700	800	900	1000	1100	1200	1300	1400
7	50	100	150	200	250	300	350	400	450	500
8	1000	900	800	700	600	500	400	300	200	100
9	2000	2500	3000	3500	4000	4500	5000	4500	4000	3500

DEMAND PROJECTIONS

The following demand is a blended demand of forecasts and orders. The periods are five-day, eight-hour periods.

An attempt has been made to illustrate that sales demand trends have little or no relationship to the resources required to produce the products, and therefore the product families.

The gross demand is as shown in Table 8.3, per item, per period.

This chapter has set up the production characteristics of the planning example. Following sections will utilize the data to exercise the logic used for top-level planning.

PART FOUR

PRODUCTION PLANNING AT A FAMILY LEVEL

CHAPTER NINE

DEVELOPMENT OF THE FAMILY PLAN

THE PURPOSE OF FAMILY LEVEL PLANNING

Part four of the book illustrates the logic necessary to perform family level production planning. The intent is to determine, at a very high level, how production can be managed to meet customer demands. As plans are completed at each level, they are fed to lower levels where the horizons grow shorter and the amount of detail increases. This level is very general, with the intent of determining if it appears that demand requirements can be met across the total horizon.

The production planning process is accomplished by performing the following steps:

Smooth the production plan for the horizon periods as necessary to meet item demands.

Develop hedge (build-ahead) inventories where necessary, but never strictly for the purpose of smoothing production.

Determine the feasibility of meeting the production plans by testing them (at a general level) against available resources.

After aggregation to a family level, required resources are tested against available facility resources to determine if the demand plan can be met. Where it cannot be met, hedge-inventory planning is performed by making build-ahead adjustments to the demand plan.

The resultant quantities per period are the production plans, and at an item level, are passed to master schedule planning.

TABLE 9.1 Item Build Time by Facility

Facility	Item	Item Run Time (min)	1	2	3	4	5	6	7	8	9
							End Items				
W/C 1	50	0.096	0.10	0.10	0.10	0.10	0.10				
W/C 2	6	1.080						1.08			
	7	1.080							1.08		
	8	0.840								0.84	
	9	0.600									0.60
W/C 3	51	4.200	4.20	4.20							
	56	4.200			4.20						
W/C 4	1	6.000	6.00								
	2	6.000		6.00							
	3	6.000			6.00						
Line 1	31	0.320				0.32	0.32				
	32	0.320						0.32	0.32	0.32	
Line 2	42	0.480				0.48	0.48				
Line 3	4	0.240				0.24					
	5	0.960					0.96				
Line 4	13	0.960						0.96	0.96	0.96	
Line 5	16	0.048						0.05	0.05	0.05	

Note: The item run times have been rounded when they are listed under the end items.

TABLE 9.2 Profile Calculations

Product Family	End Item					Run Time Requirements				
		W/C 1	W/C 2	W/C 3	W/C 4	Line 1	Line 2	Line 3	Line 4	Line 5
A	1	0.10		4.20	6.00					
	2	0.10		4.20	6.00					
	3	0.10		4.20	6.00					
	Total	0.30		12.60	18.00					
	Profile	0.10		4.20	6.00					
B	4					0.32	0.48	0.24		
	5					0.32	0.48	0.96		
	Total					0.64	0.96	1.20		
	Profile					0.32	0.48	0.60		
C	6		1.08			0.32			0.96	0.05
	7		1.08			0.32			0.96	0.05
	8		0.84			0.32			0.96	0.05
	Total		3.00			0.96			2.88	0.15
	Profile		1.00			0.32			0.96	0.05
D	9		0.60							
	Total		0.60							
	Profile		0.60							

NOTE: The above profiles are averaged quantities to the family level.

BASIC RUN TIME REQUIREMENTS

Table 9.1 illustrates the facility time required to produce the end items. For example, building an item 1 requires 0.10 minutes of W/C 1 (work center 1), 4.20 minutes of W/C 3, and 6.00 minutes of W/C 4.

FAMILY RESOURCE PROFILES

The items in a family consume about the same amount of the same resources. Therefore, the basic end item run-time requirements per facility can be accumulated to the family level and then averaged (Tables 9.2 and 9.3). The resultant family resource profile shows, on the average, the facility requirements that are necessary to produce one of any item in a product family.

AGGREGATION

Net demand (Table 9.4) is calculated by item, prior to aggregation, since it would be unrealistic to plan for the production of an item to meet a gross demand plan if a large quantity of the item was already in the warehouse.

The simple aggregation example shown in Tables 9.5 and 9.6 does not include a previous production plan, or previous hedge inventories, which in a real situation would be necessary to prevent excessive production plan nervousness from one period to the next.

If there had been a previous production plan at the family level, it would have been tested against the above demand plan, and if the demand could have been satisfied with that production plan (without over producing), that previous production plan would become the current production plan.

TABLE 9.3 A Summary of the Family Resource Profiles

Product Family	W/C 1	W/C 2	W/C 3	W/C 4	Line 1	Line 2	Line 3	Line 4	Line 5
			Run Time Requirements						
A	0.10		4.20	6.00					
B	0.10				0.32	0.48	0.60		
C		1.00			0.32			0.96	0.05
D		0.60							

TABLE 9.4 Calculation of the Net Demand

Item		Period									
		1	2	3	4	5	6	7	8	9	10
1	Gross Demand	10	30	50	80	70	60	50	40	30	20
	On hand Inventory	5									
	Net Demand	5	30	50	80	70	60	50	40	30	20
2	Gross Demand	200	200	200	200	200	200	200	200	200	200
	On hand Inventory	10									
	Net Demand	190	200	200	200	200	200	200	200	200	200
3	Gross Demand	140	130	120	100	120	140	160	180	200	220
	On hand Inventory	15									
	Net Demand	125	130	120	100	120	140	160	180	200	220
4	Gross Demand	1000	1100	1200	1300	1400	1500	1600	1700	1800	1900
	On hand Inventory	0									
	Net Demand	1000	1100	1200	1300	1400	1500	1600	1700	1800	1900
5	Gross Demand	2000	1900	1800	1700	1600	1400	1300	1200	1100	1000
	On hand Inventory	0									
	Net Demand	2000	1900	1800	1700	1600	1400	1300	1200	1100	1000

TABLE 9.4 *Continued*

	Item						Period					
			1	2	3	4	5	6	7	8	9	10
Gross Demand	6		500	600	700	800	900	1000	1100	1200	1300	1400
On hand Inventory			100									
Net Demand			400	600	700	800	900	1000	1100	1200	1300	1400
Gross Demand	7		50	100	150	200	250	300	350	400	450	500
On hand Inventory			50	100								
Net Demand			0	0	150	200	250	300	350	400	450	500
Gross Demand	8		1000	900	800	700	600	500	400	300	200	100
On hand Inventory			250									
Net Demand			750	900	800	700	600	500	400	300	200	100
Gross Demand	9		2000	2500	3000	3500	4000	4500	5000	4500	4000	3500
On hand Inventory			100									
Net Demand			1900	2500	3000	3500	4000	4500	5000	4500	4000	3500

TABLE 9.5 The Aggregation of Net Demands to the Product Family Level

Family	Item					Periods					
		1	2	3	4	5	6	7	8	9	10
A	1	5	30	50	80	70	60	50	40	30	20
	2	190	200	200	200	200	200	200	200	200	200
	3	125	130	120	100	120	140	160	180	200	220
	Total	320	360	370	380	390	400	410	420	430	440
B	4	1000	1100	1200	1300	1400	1500	1600	1700	1800	1900
	5	2000	1900	1800	1700	1600	1400	1300	1200	1100	1000
	Total	3000	3000	3000	3000	3000	2900	2900	2900	2900	2900
C	6	400	600	700	800	900	1000	1100	1200	1300	1400
	7	0	0	150	200	250	300	350	400	450	500
	8	750	900	800	700	600	500	400	300	200	100
	Total	1150	1500	1650	1700	1750	1800	1850	1900	1950	2000
D	9	1900	2500	3000	3500	4000	4500	5000	4500	4000	3500
	Total	1900	2500	3000	3500	4000	4500	5000	4500	4000	3500

TABLE 9.6   Summary of the Aggregated Product Family Net Demand Plans

Family	Period									
	1	2	3	4	5	6	7	8	9	10
A	320	360	370	380	390	400	410	420	430	440
B	3000	3000	3000	3000	3000	2900	2900	2900	2900	2900
C	1150	1500	1650	1700	1750	1800	1850	1900	1950	2000
D	1900	2500	3000	3500	4000	4500	5000	4500	4000	3500

CHAPTER TEN

WORKING WITH THE FAMILY PRODUCTION PLAN

REQUIRED RESOURCES FOR THE PRODUCTION PLAN

For each family, the resources necessary to produce one of any item in the family must be extended by the actual family quantities per period as shown in Table 10.1.

REQUIRED FAMILY RESOURCES SUMMARY

The resources required by facility are accumulated as shown in Table 10.2.

RESOURCE REQUIREMENTS TESTING

The purpose of this section is to test the required facility resources against the available facility resources. If the cumulative variance goes negative at any point, the plan cannot be met and must be reevaluated. If the cumulative variance decreases from the previous period, build-ahead production may be necessary.

For the example in Table 10.3, all times are in minutes, reflecting the number of minutes required or available in a five-day (eight-hour day) scheduling period.

Note that some examples of possible messages have been illustrated.

The following conclusions can be drawn from Table 10.3.

TABLE 10.1   Calculation of Required Family Resources

Family A

Family	Period									
	1	2	3	4	5	6	7	8	9	10

Family A net demand
| | 320 | 360 | 370 | 380 | 390 | 400 | 410 | 420 | 430 | 440 |

× the W/C 1 run time requirement of 0.10 minutes = :
| | 32 | 36 | 37 | 38 | 39 | 40 | 41 | 42 | 43 | 44 |

× the W/C 3 run time requirement of 4.20 minutes = :
| | 1344 | 1512 | 1554 | 1596 | 1638 | 1680 | 1722 | 1764 | 1806 | 1848 |

× the W/C 4 run time requirement of 6.00 minutes = :
| | 1920 | 2160 | 2220 | 2280 | 2340 | 2400 | 2460 | 2520 | 2580 | 2640 |

Family B

Family	Period									
	1	2	3	4	5	6	7	8	9	10

Family B net demand
| | 3000 | 3000 | 3000 | 3000 | 3000 | 2900 | 2900 | 2900 | 2900 | 2900 |

× the W/C 1 run time requirement of 0.10 minutes = :
| | 300 | 300 | 300 | 300 | 300 | 290 | 290 | 290 | 290 | 290 |

× the line 1 run time requirement of 0.32 minutes = :
| | 960 | 960 | 960 | 960 | 960 | 928 | 928 | 928 | 928 | 928 |

× the line 2 run time requirement of 0.48 minutes = :
| | 1440 | 1440 | 1440 | 1440 | 1440 | 1392 | 1392 | 1392 | 1392 | 1392 |

× the line 3 run time requirement of 0.60 minutes = :
| | 1800 | 1800 | 1800 | 1800 | 1800 | 1740 | 1740 | 1740 | 1740 | 1740 |

Family C

Family	Period									
	1	2	3	4	5	6	7	8	9	10

Family C net demand
| | 1150 | 1500 | 1650 | 1700 | 1750 | 1800 | 1850 | 1900 | 1950 | 2000 |

× the W/C 2 run time requirement of 1.00 minutes = :
| | 1150 | 1500 | 1650 | 1700 | 1750 | 1800 | 1850 | 1900 | 1950 | 2000 |

× the line 1 run time requirement of 0.32 minutes = :
| | 368 | 480 | 528 | 544 | 560 | 576 | 592 | 608 | 624 | 640 |

× the line 4 run time requirement of 0.96 minutes = :
| | 1104 | 1440 | 1584 | 1632 | 1680 | 1728 | 1776 | 1824 | 1872 | 1920 |

× the line 5 run time requirement of 0.05 minutes = :
| | 58 | 75 | 83 | 85 | 88 | 90 | 93 | 95 | 98 | 100 |

TABLE 10.1   (Continued)

Family D

Family	1	2	3	4	5	6	7	8	9	10
					Period					

Family D net demand

	1900	2500	3000	3500	4000	4500	5000	4500	4000	3500

× the W/C 2 run time requirement of 0.60 minutes = :

	1140	1500	1800	2100	2400	2700	3000	2700	2400	2100

All production plans can be met with the existing facilities since there are no negative cumulative variances.

Some production plans may require adjustment with build aheads due to a decrease in the variance from one period to the next (refer to the warnings in the tables).

FAMILY MANAGEMENT

The previous resource requirements testing indicated potential problems with work centers 2 and 4.

Work center 2 supports families C and D. Although family C has an increasing requirement across time, family D has a requirement spike in the same period that was flagged (period 7) as a potential problem area during resource testing. The production planner elects to adjust family D.

Family D contains only one end item, which is number 9. The adjustment process adjusts the period quantities for a selected item (within the family), which automatically adjusts the family period quantities. For clarification, both adjustment steps are shown in Table 10.4.

Note that the adjustments have been made based on what the production planner thinks will solve the problem. This may not always be correct where many items pass through a particular facility. In a real life situation, another iteration of resource testing would be required after adjustments are made.

Work center 4 supports only family A. Family A contains end item numbers 1, 2, and 3. Item 1 has a seasonal pattern that is on the decrease when the facility shortage occurs. Item 2 has a nearly flat requirement pattern. Item 3 has an increasing requirement pattern when the facility shortage occurs. The production planner elects to adjust item 3 for some build ahead.

The adjustment process adjusts the period quantities for a selected item (within the family), which automatically adjusts the family period quantities. For clarification, both adjustment steps are shown in Table 10.5.

Production Planning at a Family Level

TABLE 10.2 Summary of Required Resources by Facility

Facility	Period									
	1	2	3	4	5	6	7	8	9	10
W/C 1 Family										
A	32	36	37	38	39	40	41	42	43	44
B	300	300	300	300	300	290	290	290	290	290
Total	332	336	337	338	339	330	331	332	333	334
W/C 2 Family										
C	1150	1500	1650	1700	1750	1800	1850	1900	1950	2000
D	1140	1500	1800	2100	2400	2700	3000	2700	2400	2100
Total	2290	3000	3450	3800	4150	4500	4850	4600	4350	4100
W/C 3 Family										
A	1344	1512	1554	1596	1638	1680	1722	1764	1806	1848
W/C 4 Family										
A	1920	2160	2220	2280	2340	2400	2460	2520	2580	2640
Line 1 Family										
B	960	960	960	960	960	928	928	928	928	928
C	368	480	528	544	560	576	592	608	624	640
Total	1328	1440	1488	1504	1520	1504	1520	1536	1550	1568
Line 2 Family										
B	1440	1440	1440	1440	1440	1392	1392	1392	1392	1392
Line 3 Family										
B	1800	1800	1800	1800	1800	1740	1740	1740	1740	1740
Line 4 Family										
C	1104	1440	1584	1632	1680	1728	1776	1824	1872	1920
Line 5 Family										
C	58	75	83	85	88	90	93	95	98	100

TABLE 10.3 Testing Required Resources Against Available

W/C 1

					Period					
	1	2	3	4	5	6	7	8	9	10
Required	332	336	337	338	339	330	331	332	333	334
Available	2,400	2,400	2,400	2,400	2,400	2,400	2,400	2,400	2,400	2,400
Variance	2,068	2,064	2,063	2,062	2,061	2,070	2,069	2,068	2,067	2,066
Cumulative	2,068	4,132	6,195	8,257	10,318	12,388	14,457	16,525	18,592	20,658

Caution: This facility has a utilization of 14 % .

W/C 2

					Period					
	1	2	3	4	5	6	7	8	9	10
Required	2290	3000	3450	3800	4150	4500	4850	4600	4350	4100
Available	4800	4800	4800	4800	4800	4800	4800	4800	4800	4800
Variance	2510	1800	1350	1000	650	300	−50	200	450	700
Cumulative	2510	4310	5660	6660	7310	7610	7560	7760	8210	8910

Warning: The period 7 variance is less than the period 6 variance and may require a build-ahead.

W/C 3

					Period					
	1	2	3	4	5	6	7	8	9	10
Required	1344	1512	1554	1596	1638	1680	1722	1764	1806	1848
Available	2400	2400	2400	2400	2400	2400	2400	2400	2400	2400
Variance	1056	888	846	804	762	720	678	636	594	552
Cumulative	1056	1944	2790	3594	4356	5076	5754	6390	6984	7536

TABLE 10.3 (Continued)

W/C 4					Period					
	1	2	3	4	5	6	7	8	9	10
Required	1920	2160	2220	2280	2340	2400	2460	2520	2580	2640
Available	2400	2400	2400	2400	2400	2400	2400	2400	2400	2400
Variance	480	240	180	120	60	0	−60	−120	−180	−240
Cumulative	480	720	900	1020	1080	1080	1020	900	720	480

Warning: Periods 7–10 variances are decreasing and may require a build ahead.

Line 1					Period					
	1	2	3	4	5	6	7	8	9	10
Required	1328	1440	1488	1504	1520	1504	1520	1536	1550	1568
Available	2400	2400	2400	2400	2400	2400	2400	2400	2400	2400
Variance	1072	960	912	896	880	896	880	864	850	832
Cumulative	1072	2032	2944	3840	4720	5616	6496	7360	8210	9042

Line 2					Period					
	1	2	3	4	5	6	7	8	9	10
Required	1440	1440	1440	1440	1440	1392	1392	1392	1392	1392
Available	2400	2400	2400	2400	2400	2400	2400	2400	2400	2400
Variance	960	960	960	960	960	1008	1008	1008	1008	1008
Cumulative	960	1920	2880	3840	4800	5808	6816	7824	8832	9840

Period

Line 3	1	2	3	4	5	6	7	8	9	10
Required	1800	1800	1800	1800	1800	1740	1740	1740	1740	1740
Available	2400	2400	2400	2400	2400	2400	2400	2400	2400	2400
Variance	600	600	600	600	600	660	660	660	660	660
Cumulative	600	1200	1800	2400	3000	3660	4320	4980	5640	6300

Period

Line 4	1	2	3	4	5	6	7	8	9	10
Required	1104	1440	1584	1632	1680	1728	1776	1824	1872	1920
Available	2400	2400	2400	2400	2400	2400	2400	2400	2400	2400
Variance	1296	960	816	768	720	672	624	576	528	480
Cumulative	1296	2256	3072	3840	4560	5232	5856	6432	6960	7440

Period

Line 5	1	2	3	4	5	6	7	8	9	10
Required	58	75	83	85	88	90	93	95	98	100
Available	2,400	2,400	2,400	2,400	2,400	2,400	2,400	2,400	2,400	2,400
Variance	2,342	2,325	2,317	2,315	2,312	2,310	2,307	2,305	2,302	2,300
Cumulative	2,342	4,667	6,984	9,299	11,611	13,921	16,228	18,533	20,835	23,135

Caution: This facility has a utilization of 4%.

109

TABLE 10.4   Family D Management

The Initial Plan

					Period					
	1	2	3	4	5	6	7	8	9	10
Item 9	1900	2500	3000	3500	4000	4500	5000	4500	4000	3500
Family D	1900	2500	3000	3500	4000	4500	5000	4500	4000	3500

Note: The data source for item 9 is Table 9.4 and for family D is Table 9.5.

The Adjustments

Resource testing showed a shortage of 50 minutes in period 7. Item 9 consumed 0.60 minutes of W/C 2 per item produced. The adjustment should be equal to at least 85 (50 divided by 0.60 = 83.3).

					Period					
	1	2	3	4	5	6	7	8	9	10
Adjustments						+ 85	− 85			

The Resultant Plan

					Period					
	1	2	3	4	5	6	7	8	9	10
Item 9	1900	2500	3000	3500	4000	4585	4915	4500	4000	3500
Family D	1900	2500	3000	3500	4000	4585	4915	4500	4000	3500

TABLE 10.5   Family A Management

The Initial Plan

					Period					
	1	2	3	4	5	6	7	8	9	10
Item 3	125	130	120	100	120	140	160	180	200	220
Family A	320	360	370	380	390	400	410	420	430	440

Note: The data source for item 3 is Table 9.4 and for family A is Table 9.5.

The Adjustments

Resource testing showed a shortage of 60 minutes in period 7 and an excess of 60 minutes in period 5. Item 3 consumed 6.00 minutes of W/C 4 per item produced. The adjustment should be equal to at least 10 (60 divided by 6.00 = 10.0).

					Period					
	1	2	3	4	5	6	7	8	9	10
Adjustments					+ 10		− 10			

The same logic applies to periods 8, 9 and 10, which results in the following total adjustment picture.

					Period					
	1	2	3	4	5	6	7	8	9	10
Adjustments		+ 40	+ 30	+ 20	+ 10		− 10	− 20	− 30	− 40

The Resultant Plan

					Period					
	1	2	3	4	5	6	7	8	9	10
Item 3	125	170	150	120	130	140	150	160	170	180
Family A	320	400	400	400	400	400	400	400	400	400

TABLE 10.6   Summary of the Item Production Plans

	Period									
Item	1	2	3	4	5	6	7	8	9	10
1	5	30	50	80	70	60	50	40	30	20
2	190	200	200	200	200	200	200	200	200	200
3	125	170	150	120	130	140	150	160	170	180
4	1000	1100	1200	1300	1400	1500	1600	1700	1800	1900
5	2000	1900	1800	1700	1600	1400	1300	1200	1100	1000
6	400	600	700	800	900	1000	1100	1200	1300	1400
7	0	0	150	200	250	300	350	400	450	500
8	750	900	800	700	600	500	400	300	200	100
9	1900	2500	3000	3500	4000	4585	4915	4500	4000	3500

Note: The above data (in this case study) is derived from the net demands and the adjusted items under family management. In an actual situation, the previous period's item production plans would also be an input factor.

ITEM PRODUCTION PLANS

For summarization purposes, the resultant item production plans that are passed to master schedule planning are shown in Table 10.6. Note that the derivation of these item plans did not require disaggregation of the family plans as any adjustments applied to both sets of plans.

These item production plans represent how product can be built to satisfy demand within facility constraints and without building up excessive inventories. The inventories that are build aheads are hedge inventories, and their balances must be maintained in a separate set of records from the available on-hand inventory balances.

This concludes the section on family production planning. In the next section we take the item production plans that were developed and perform master schedule planning at an item (not family) level.

PART FIVE

MASTER SCHEDULE PLANNING AT AN ITEM LEVEL

CHAPTER ELEVEN

BASIC MASTER SCHEDULE PLANNING

PLANNING FOR LOT SIZED ITEMS

The following items (with their production plan requirements) are planned for production based on lot sizing controls at the indicated facilities. The data has been extracted from the previous chapter's Summary of the Item Production Plans.

Planning will be based on the indicated lot sized quantity, or some multiple of that quantity. Note that items 4 and 5 are not included, as these items are not produced based on lot sizes. They are produced based on a rate schedule of a certain quantity per period.

Table 11.1 shows the item production plans that apply to lot sized items. Each of the 10 periods in the above table consists of five days. These five-day periods have to be converted to one-day periods for master schedule planning. If the planning period in production planning had been monthly, we might have reduced it to weekly (five-day periods) for master schedule planning. In the example, however, we have been planning in weekly periods, so in this next lower level of planning, we will reduce the weekly period to a daily period.

The production plan for a week (Table 11.1) is distributed across the five days that are represented. This is called the distributed production plan.

Planned lot size quantities to be received on specific dates (Planned Receipt) are based on what was available from the previous period (Previous period available), minus what is required to support the production plan for

TABLE 11.1 Planned Production Quantities for Lot Sized Items

Item	W/C	Lot Size	Period									
			1	2	3	4	5	6	7	8	9	10
1	4	10	5	30	50	80	70	60	50	40	30	20
2	4	20	190	200	200	200	200	200	200	200	200	200
3	4	30	125	170	150	120	130	140	150	160	170	180
6	6	200	400	600	700	800	900	1000	1100	1200	1300	1400
7	6	300	0	0	150	200	250	300	350	400	450	500
8	6	500	750	900	800	700	600	500	400	300	200	100
9	6	200	1900	2500	3000	3500	4000	4585	4915	4500	4000	3500

this period. If there was enough inventory remaining from the previous period, then a planned receipt is not necessary.

In the Table 11.2, the weekly production plan (Production Plan) is distributed to the distributed production plan. If the previous period available is less than the distributed production plan, then a planned receipt is added. The table is used to calculate what lot sized quantity is required on what date.

We have now developed a planned receipt quantity and a date for each lot size planned item. Remember that this is still a very gross plan that has not been tested against available resources to see if it can actually be achieved.

However, before we can do resource testing, we have to also determine a gross plan for the rate controlled items, since our items are intermixed with both work centers and flow lines.

PLANNING FOR RATE CONTROLLED ITEMS

Items 4 and 5 are to be scheduled based on rate controls on line 3 to meet the previously defined Item Production Plans.

Planning (for this example) is based on the desired rate whenever possible. The minimum rate is utilized by the master scheduler only when it is necessary to keep the line operational (and the labor employed). The maximum rate is utilized to satisfy peak production requirements that cannot be satisfied by build aheads at the desired rate.

Note that other scheduling algorithms could be applied. If the rate was totally controlled by the process, then the line would run at the desired rate or not run at all. Another circumstance might be that the company could vary the daily employment of direct labor. If they could, then they would also run only at the desired rate and not be concerned with minimums and maximums.

How long a line operates under these alternative approaches is dependent upon seasonal demand periods, finished goods inventory carrying costs, and line setup costs.

The previously developed item production plans are as shown in Table 11.3.

Each of the above periods consist of five days and has to be converted to one-day periods for the distributed production plan, as was done for the lot size controlled items.

The rate parameters for these items are as shown in Table 11.4.

Table 11.4 is a simple comparison of the daily rates to the five-day (weekly) rates. The indicator of a potential planning problem is that the rates for the two items which run on the line are different. This is one of two variables. The other is that the mix quantity can change period by period. As a result, the planning logic may have to differ from lot sized item planning logic to be effective.

TABLE 11.2 Lot Sized Planning

End Item 1: The minimum lot size is 10.

	Periods in Days									
	1	2	3	4	5	6	7	8	9	10
Production plan	5					30				
+ Previous period available	0	9	8	7	6	5	9	3	7	1
+ Planned receipt	10					10		10		1
= Total available	10	9	8	7	6	15	9	3	7	11
− Distributed production plan	1	1	1	1	1	6	6	6	6	6
= Current period available	9	8	7	6	5	9	3	7	1	5

	Periods in Days									
	11	12	13	14	15	16	17	18	19	20
Production plan	50					80				
+ Previous period available	5	5	5	5	5	5	9	3	7	1
+ Planned receipt	10	10	10	10	10	20	10	20	10	20
= Total available	15	15	15	15	15	25	19	23	17	21
− Distributed production plan	10	10	10	10	10	16	16	16	16	16
= Current period available	5	5	5	5	5	9	3	7	1	5

Periods in Days

	21	22	23	24	25	26	27	28	29	30
Production plan	70					60				
+ Previous period available	5	1	7	3	9	5	3	1	9	7
+ Planned receipt	10	20	10	20	10	10	10	20	10	10
= Total available	15	21	17	23	19	15	13	21	19	17
− Distributed production plan	14	14	14	14	14	12	12	12	12	12
= Current period available	1	7	3	9	5	3	1	9	7	5

Periods in Days

	31	32	33	34	35	36	37	38	39	40
Production plan	50					40				
+ Previous period available	5	5	5	5	5	5	7	9	1	3
+ Planned receipt	10	10	10	20	10	10	10		10	10
= Total available	15	15	15	15	15	15	17	9	11	13
− Distributed production plan	10	10	10	10	10	8	8	8	8	8
= Current period available	5	5	5	5	5	7	9	1	3	5

Periods in Days

	41	42	43	44	45	46	47	48	49	50
Production plan	30					20				
+ Previous period available	5	9	3	7	1	5	11	7	3	9
+ Planned receipt	10		10		10	10			10	
= Total available	15	9	13	7	11	15	11	7	13	9
− Distributed production plan	6	6	6	6	6	4	4	4	4	4
= Current period available	9	3	7	1	5	11	7	3	9	5

TABLE 11.2 (Continued)

End Item 2: The minimum lot size is 20.

					Periods in Days					
	1	2	3	4	5	6	7	8	9	10
Production plan	190					200				
+ Previous period available	0	2	4	6	8	10	10	10	10	10
+ Planned receipt	40	40	40	40	40	40	40	40	40	40
= Total available	40	42	44	46	48	50	50	50	50	50
− Distributed production plan	38	38	38	38	38	40	40	40	40	40
= Current period available	2	4	6	8	10	10	10	10	10	10

					Periods in Days					
	11	12	13	14	15	16	17	18	19	20
Production plan	200					200				
+ Previous period available	10	10	10	10	10	10	10	10	10	10
+ Planned receipt	40	40	40	40	40	40	40	40	40	40
= Total available	50	50	50	50	50	50	50	50	50	50
− Distributed production plan	40	40	40	40	40	40	40	40	40	40
= Current period available	10	10	10	10	10	10	10	10	10	10

Periods in Days

	21	22	23	24	25	26	27	28	29	30
Production plan						200				
+ Previous period available	10	10	10	10	10	10	10	10	10	10
+ Planned receipt	40	40	40	40	40	40	40	40	40	40
= Total available	50	50	50	50	50	50	50	50	50	50
− Distributed production plan	40	40	40	40	40	40	40	40	40	40
= Current period available	10	10	10	10	10	10	10	10	10	10

Periods in Days

	31	32	33	34	35	36	37	38	39	40
Production plan						200				
+ Previous period available	10	10	10	10	10	10	10	10	10	10
+ Planned receipt	40	40	40	40	40	40	40	40	40	40
= Total available	50	50	50	50	50	50	50	50	50	50
− Distributed production plan	40	40	40	40	40	40	40	40	40	40
= Current period available	10	10	10	10	10	10	10	10	10	10

Periods in Days

	41	42	43	44	45	46	47	48	49	50
Production plan						200				
+ Previous period available	10	10	10	10	10	10	10	10	10	10
+ Planned receipt	40	40	40	40	40	40	40	40	40	40
= Total available	50	50	50	50	50	50	50	50	50	50
− Distributed production plan	40	40	40	40	40	40	40	40	40	40
= Current period available	10	10	10	10	10	10	10	10	10	10

TABLE 11.2 (Continued)

End Item 3: The minimum lot size is 30.

					Periods in Days					
	1	2	3	4	5	6	7	8	9	10
Production plan	125					170				
+ Previous period available	0	5	10	15	20	25	21	17	13	9
+ Planned receipt	30	30	30	30	30	30	30	30	30	30
= Total available	30	35	40	45	50	55	51	47	43	39
− Distributed production plan	25	25	25	25	25	34	34	34	34	34
= Current period available	5	10	15	20	25	21	17	13	9	5

					Periods in Days					
	11	12	13	14	15	16	17	18	19	20
Production plan	150					120				
+ Previous period available	5	5	5	5	5	5	11	17	23	29
+ Planned receipt	30	30	30	30	30	30	30	30	30	
= Total available	35	35	35	35	35	35	41	47	53	29
− Distributed production plan	30	30	30	30	30	24	24	24	24	24
= Current period available	5	5	5	5	5	11	17	23	29	5

	21	22	23	24	25	26	27	28	29	30
Production plan	130					140				
+ Previous period available	5	9	13	17	21	25	27	29	1	3
+ Planned receipt	30	30	30	30	30	30	30		30	30
= Total available	35	39	43	47	51	55	57	29	31	33
− Distributed production plan	26	26	26	26	26	28	28	28	28	28
= Current period available	9	13	17	21	25	27	29	1	3	5

	31	32	33	34	35	36	37	38	39	40
Production plan	150					160				
+ Previous period available	5	5	5	5	5	5	3	1	29	27
+ Planned receipt	30	30	30	30	30	30	30	60	30	30
= Total available	35	35	35	35	35	35	33	61	59	57
− Distributed production plan	30	30	30	30	30	32	32	32	32	32
= Current period available	5	5	5	5	5	3	1	29	27	25

	41	42	43	44	45	46	47	48	49	50
Production plan	170					180				
+ Previous period available	25	21	17	13	9	5	29	23	17	11
+ Planned receipt	30	30	30	30	30	60	30	30	30	30
= Total available	55	51	47	43	39	65	59	53	47	41
− Distributed production plan	34	34	34	34	34	36	36	36	36	36
= Current period available	21	17	13	9	5	29	23	17	11	5

TABLE 11.2 (Continued)

End Item 6: The minimum lot size is 200.

					Periods in Days					
	1	2	3	4	5	6	7	8	9	10
Production plan	400					600				
+ Previous period available	0	120	40	160	80	0	80	160	40	120
+ Planned receipt	200		200			200	200		200	
= Total available	200	120	240	160	80	200	280	160	240	120
− Distributed production plan	80	80	80	80	80	120	120	120	120	120
= Current period available	120	40	160	80	0	80	160	40	120	0

					Periods in Days					
	11	12	13	14	15	16	17	18	19	20
Production plan	700					800				
+ Previous period available	0	60	120	180	40	100	140	180	20	60
+ Planned receipt	200	200	200		200	200	200		200	200
= Total available	200	260	320	180	240	300	340	180	220	260
− Distributed production plan	140	140	140	140	140	160	160	160	160	160
= Current period available	60	120	180	40	100	140	180	20	60	100

124

Periods in Days

	21	22	23	24	25	26	27	28	29	30
Production plan	900					1000				
+ Previous period available	100	120	140	160	180	0	0	0	0	0
+ Planned receipt	200	200	200	200	0	200	200	200	200	200
= Total available	300	320	340	360	180	200	200	200	200	200
− Distributed production plan	180	180	180	180	180	200	200	200	200	200
= Current period available	120	140	160	180	0	0	0	0	0	0

Periods in Days

	31	32	33	34	35	36	37	38	39	40
Production plan	1100					1200				
+ Previous period available	0	180	160	140	120	100	60	20	180	140
+ Planned receipt	400	200	200	200	200	200	200	400	200	200
= Total available	400	380	360	340	320	300	260	420	380	340
− Distributed production plan	220	220	220	220	220	240	240	240	240	240
= Current period available	180	160	140	120	100	60	20	180	140	100

Periods in Days

	41	42	43	44	45	46	47	48	49	50
Production plan	1300					1400				
+ Previous period available	100	40	180	120	60	0	120	40	160	80
+ Planned receipt	200	400	200	200	200	400	200	400	200	200
= Total available	300	440	380	320	260	400	320	440	360	280
− Distributed production plan	260	260	260	260	260	280	280	280	280	280
= Current period available	40	180	120	60	0	120	40	160	80	0

TABLE 11.2 (Continued)

End Item 7: The minimum lot size is 300.

	1	2	3	4	5	6	7	8	9	10
Production plan	0					0				
+ Previous period available		0	0	0	0	0	0	0	0	0
+ Planned receipt										
= Total available	0	0	0	0	0	0	0	0	0	0
− Distributed production plan	0	0	0	0	0	0	0	0	0	0
= Current period available	0	0	0	0	0	0	0	0	0	0

Periods in Days

	11	12	13	14	15	16	17	18	19	20
Production plan	150					200				
+ Previous period available	0	270	240	210	180	150	110	70	30	290
+ Planned receipt	300								300	
= Total available	300	270	240	210	180	150	110	70	330	290
− Distributed production plan	30	30	30	30	30	40	40	40	40	40
= Current period available	270	240	210	180	150	110	70	30	290	250

Periods in Days

Periods in Days

	21	22	23	24	25	26	27	28	29	30
Production plan	250					300				
+ Previous period available	250	200	150	100	50	0	240	180	120	60
+ Planned receipt						300				
= Total available	250	200	150	100	50	300	240	180	120	60
− Distributed production plan	50	50	50	50	50	60	60	60	60	60
= Current period available	200	150	100	50	0	240	180	120	60	0

Periods in Days

	31	32	33	34	35	36	37	38	39	40
Production plan	350					400				
+ Previous period available	0	230	160	90	20	250	170	90	10	230
+ Planned receipt	300				300				300	
= Total available	300	230	160	90	320	250	170	90	310	230
− Distributed production plan	70	70	70	70	70	80	80	80	80	80
= Current period available	230	160	90	20	250	170	90	10	230	150

Periods in Days

	41	42	43	44	45	46	47	48	49	50
Production plan	450					500				
+ Previous period available	150	60	270	180	90	0	200	100	0	200
+ Planned receipt		300				300			300	
= Total available	150	360	270	180	90	300	200	100	300	200
− Distributed production plan	90	90	90	90	90	100	100	100	100	100
= Current period available	60	270	180	90	0	200	100	0	200	100

TABLE 11.2 (Continued)

End Item 8: The minimum lot size is 500.

					Periods in Days					
	1	2	3	4	5	6	7	8	9	10
Production plan	750					900				
+ Previous period available	0	350	200	50	400	250	70	390	210	30
+ Planned receipt	500			500			500			500
= Total available	500	350	200	550	400	250	570	390	210	530
− Distributed production plan	150	150	150	150	150	180	180	180	180	180
= Current period available	350	200	50	400	250	70	390	210	30	350

					Periods in Days					
	11	12	13	14	15	16	17	18	19	20
Production plan	800					700				
+ Previous period available	350	190	30	370	210	50	410	270	130	490
+ Planned receipt			500			500			500	
= Total available	350	190	530	370	210	550	410	270	630	490
− Distributed production plan	160	160	160	160	160	140	140	140	140	140
= Current period available	190	30	370	210	50	410	270	130	490	350

Periods in Days

	21	22	23	24	25	26	27	28	29	30
Production plan	600					500				
+ Previous period available	350	230	110	490	370	250	150	50	450	350
+ Planned receipt			500					500		
= Total available	350	230	610	490	370	250	150	550	450	350
− Distributed production plan	120	120	120	120	120	100	100	100	100	100
= Current period available	230	110	490	370	250	150	50	450	350	250

Periods in Days

	31	32	33	34	35	36	37	38	39	40
Production plan	400					300				
+ Previous period available	250	170	90	10	430	350	290	230	170	110
+ Planned receipt				500						
= Total available	250	170	90	510	430	350	290	230	170	110
− Distributed production plan	80	80	80	80	80	60	60	60	60	60
= Current period available	170	90	10	430	350	290	230	170	110	50

Periods in Days

	41	42	43	44	45	46	47	48	49	50
Production plan	200					100				
+ Previous period available	50	10	470	430	390	350	330	310	290	270
+ Planned receipt		500								
= Total available	50	510	470	430	390	350	330	310	290	270
− Distributed production plan	40	40	40	40	40	20	20	20	20	20
= Current period available	10	470	430	390	350	330	310	290	270	250

TABLE 11.2 (Continued)

End Item 9: The minimum lot size is 200.

					Periods in Days					
	1	2	3	4	5	6	7	8	9	10
Production plan	1900					2500				
+ Previous period available	0	20	40	60	80	100	0	100	0	100
+ Planned receipt	400	400	400	400	400	400	600	400	600	400
= Total available	400	420	440	460	480	500	600	500	600	500
− Distributed production plan	380	380	380	380	380	500	500	500	500	500
= Current period available	20	40	60	80	100	0	100	0	100	0

					Periods in Days					
	11	12	13	14	15	16	17	18	19	20
Production plan	3000					3500				
+ Previous period available	0	0	0	0	0	0	100	0	100	0
+ Planned receipt	600	600	600	600	600	800	600	800	600	800
= Total available	600	600	600	600	600	800	700	800	700	800
− Distributed production plan	600	600	600	600	600	700	700	700	700	700
= Current period available	0	0	0	0	0	100	0	100	0	100

130

Periods in Days

	21	22	23	24	25	26	27	28	29	30
Production plan	4000					4585				
+ Previous period available	100	100	100	100	100	100	183	66	149	32
+ Planned receipt	800	800	800	800	800	1000	800	1000	800	1000
= Total available	900	900	900	900	900	1100	983	1066	949	1032
− Distributed production plan	800	800	800	800	800	917	917	917	917	917
= Current period available	100	100	100	100	100	183	66	149	32	115

Periods in Days

	31	32	33	34	35	36	37	38	39	40
Production plan	4915					4500				
+ Previous period available	115	132	149	166	183	0	100	200	100	200
+ Planned receipt	1000	1000	1000	1000	800	1000	1000	800	1000	800
= Total available	1115	1132	1149	1166	983	1000	1100	1000	1100	1000
− Distributed production plan	983	983	983	983	983	900	900	900	900	900
= Current period available	132	149	166	183	0	100	200	100	200	100

Periods in Days

	41	42	43	44	45	46	47	48	49	50
Production plan	4000					3500				
+ Previous period available	100	100	100	100	100	100	0	100	0	100
+ Planned receipt	800	800	800	800	800	600	800	600	800	600
= Total available	900	900	900	900	900	700	800	700	800	700
− Distributed production plan	800	800	800	800	800	700	700	700	700	700
= Current period available	100	100	100	100	100	0	100	0	100	0

TABLE 11.3   Planned Production Quantities for Rate Controlled Items

		Period									
Item	Line	1	2	3	4	5	6	7	8	9	10
4	3	1000	1100	1200	1300	1400	1500	1600	1700	1800	1900
5	3	2000	1900	1800	1700	1600	1400	1300	1200	1100	1000

TABLE 11.4   Rate Comparisons

Rates per Day

Line	Item	Minimum Rate	Desired Rate	Maximum Rate
3	4	1,800	2,000 (1)	2,200
	5	450	500 (1)	550

Rates per Five-Day Period

Line	Item	Maximum Rate	Desired Rate	Maximum Rate
3	4	9,000	10,000 (2)	11,000
	5	2,250	2,500 (2)	2,750

Notes:

(1) 2000 item 4s, or 500 item 5s, or some quantity of a mixture of items 4 and 5 (the quantity being dependent on the mixture) is the desired daily rate for the line.

(2) Note (1) should be considered to be extended for the five-day planning periods.

THE CALCULATION FOR INTERSPERSED ITEMS RUN ON THE SAME LINE

The above table has indicated that 2000 item 4s, or 500 item 5s, or some quantity of a mixture of items 4 and 5 (the quantity being dependent on the mixture) determines the desired daily rate for the line.

The scheduling of line 3 is based on the combination of item 4 and item 5 production plan requirements. The constraints are the minimum, desired, and maximum rates for the line, regardless of item mix.

This means that the production plan requirements must be evaluated against the line capacities to determine the correct rate to run the line. The capacity constraint that will be utilized is minutes of time.

The basic calculation for interspersed items on one line is as follows:

(The planned quantity of item A × the required resource for 1 item A)

+ (the planned quantity of item N × the required resource for 1 item N)

= the total required resource

The calculated total required resource (of minutes) is then compared to the available minimum, desired, and maximum resources, and adjustments are made as necessary. Note that the base is the number of minutes in a day (480).

This is a significant difference in the planning logic of a job shop versus a flow shop. Quantities in a job shop were planned as to when a lot sized quantity was required. In this flow shop example, since the two products (that run on the same line) have different run (or build) times, the item quantities have to be converted from the number of "eaches" to the amount of required "times," which we will measure in minutes.

To indicate that a minimum rate is in effect, a 10% decrease in available resource (432 minutes) could be reflected. This means that what could normally be produced in 432 minutes, actually would take 480 minutes to produce.

At the conclusion of this resource balancing process, the quantities will be converted back to item production quantities.

The potential plus and minus rate variations are shown in Table 11.5. Note that in Table 11.5, we are saying that to produce at a slower rate means it takes longer to produce any one item. It can be viewed that way or viewed

TABLE 11.5   Rate Variations

Item Rate Variations by Minutes to Produce One Item

Line	Item	Minimum Rate	Desired Rate	Maximum Rate
		−10%		+10%
3	4	0.27	0.24	0.22
	5	1.07	0.96	0.87

Line Rate Variations by Minutes that Are Available for Production

Line	Item	Minimum Rate	Desired Rate	Maximum Rate
		−10%		+10%
3		432	480	528

as if the day of 480 minutes actually consisted of only 432 minutes. Either approach will provide the same results.

Another approach is to say that it always takes X time to produce an item. If my required production time is less than the time I have available (for the rate at which I wish to run), then I will produce hedge or build-ahead inventory. This can also be viewed as building hedge time. That is, we are consuming minutes now that we may have consumed in the future if we had run at a slower rate now. This case study example will utilize this latter approach.

RATE PLANNING BY RESOURCE

Table 11.6 shows the weekly production plan distributed to a daily production plan. This planned production quantity is then extended by the minutes of required production time for each item. The rate used in this initial calculation is the desired rate, which, if possible, is how we would like to run the line. Table 11.6 has provided us with individual daily minutes required for the production of each item necessary to satisfy the item production plan.

COMBINING THE FACILITY RESOURCE TIME REQUIREMENTS FOR RATE SCHEDULED ITEMS ON THE SAME FACILITY

We now need to total the individual item time requirements and calculate the mix percentage (Table 11.7). By knowing the mix percentage, we will be able to vary the line rate in future calculations and see the impact on the amount of time required to produce each item.

Note that in Table 11.7, the mix percentage has changed in every new weekly planning period. In theory, it could be managed so that it would be allowed to change daily or even hourly. In practice, most companies are content to set up a line to run at a certain rate with a specified mix of product for a number of days or weeks.

MASTER SCHEDULING THE RATE CONTROLLED ITEMS

The allowable rate variations for the line in daily minutes were defined as follows:

Minimum = 432

Desired = 480

Maximum = 528

TABLE 11.6 Rate Planning

End Item 4: The resource required to produce a single item is 0.24 minutes

	Periods in Days									
	1	2	3	4	5	6	7	8	9	10
Production plan	1000					1100				
Distribution production plan	200	200	200	200	200	220	220	220	220	220
× 0.24 minutes =	48	48	48	48	48	53	53	53	53	53

	Periods in Days									
	11	12	13	14	15	16	17	18	19	20
Production plan	1200					1300				
Distribution production plan	240	240	240	240	240	260	260	260	260	260
× 0.24 minutes =	58	58	58	58	58	62	62	62	62	62

	Periods in Days									
	21	22	23	24	25	26	27	28	29	30
Production plan	1400					1500				
Distribution production plan	280	280	280	280	280	300	300	300	300	300
× 0.24 minutes =	67	67	67	67	67	72	72	72	72	72

TABLE 11.6 (Continued)

					Periods in Days					
	31	32	33	34	35	36	37	38	39	40
Production plan	1600					1700				
Distribution production plan	320	320	320	320	320	340	340	340	340	340
× 0.24 minutes =	77	77	77	77	77	82	82	82	82	82

					Periods in Days					
	41	42	43	44	45	46	47	48	49	50
Production plan	1800					1900				
Distribution production plan	360	360	360	360	360	380	380	380	380	380
× 0.24 minutes =	86	86	86	86	86	91	91	91	91	91

End Item 5: The resource required to produce a single item is 0.96 minutes

					Periods in Days					
	1	2	3	4	5	6	7	8	9	10
Production plan	2000					1900				
Distribution production plan	400	400	400	400	400	380	380	380	380	380
× 0.96 minutes =	384	384	384	384	384	365	365	365	365	365

Periods in Days

	11	12	13	14	15	16	17	18	19	20
Production plan	1800					1700				
Distribution production plan	360	360	360	360	360	340	340	340	340	340
× 0.96 minutes =	346	346	346	346	346	326	326	326	326	326

Periods in Days

	21	22	23	24	25	26	27	28	29	30
Production plan	1600					1400				
Distribution production plan	320	320	320	320	320	280	280	280	280	280
× 0.96 minutes =	307	307	307	307	307	269	269	269	269	269

Periods in Days

	31	32	33	34	35	36	37	38	39	40
Production plan	1300					1200				
Distribution production plan	260	260	260	260	260	240	240	240	240	240
× 0.96 minutes =	250	250	250	250	250	230	230	230	230	230

Periods in Days

	41	42	43	44	45	46	47	48	49	50
Production plan	1100					1000				
Distribution production plan	220	220	220	220	220	200	200	200	200	200
× 0.96 minutes =	211	211	211	211	211	192	192	192	192	192

TABLE 11.7 Calculation of the Mix Percentage of Multiple Items on the Same Line

	Periods in Days									
	1	2	3	4	5	6	7	8	9	10
Item 4 required time	48	48	48	48	48	53	53	53	53	53
Item 5 required time	384	384	384	384	384	365	365	365	365	365
Total	432	432	432	432	432	418	418	418	418	418
Item 4 %	0.11					0.13				
Item 5 %	0.80					0.87				

	Periods in Days									
	11	12	13	14	15	16	17	18	19	20
Item 4 required time	58	58	58	58	58	62	62	62	62	62
Item 5 required time	346	346	346	346	346	326	326	326	326	326
Total	404	404	404	404	404	388	388	388	388	388
Item 4 %	0.14					0.16				
Item 5 %	0.86					0.84				

	Periods in Days									
	21	22	23	24	25	26	27	28	29	30
Item 4 required time	67	67	67	67	67	72	72	72	72	72
Item 5 required time	307	307	307	307	307	269	269	269	269	269
Total	374	374	374	374	374	341	341	341	341	341
Item 4 %	0.17					0.21				
Item 5 %	0.83					0.79				

Periods in Days

	31	32	33	34	35	36	37	38	39	40
Item 4 required time	77	77	77	77	77	82	82	82	82	82
Item 5 required time	250	250	250	250	250	230	230	230	230	230
Total	327	327	327	327	327	312	312	312	312	312
Item 4 %	0.24					0.26				
Item 5 %	0.76					0.74				

Periods in Days

	41	42	43	44	45	46	47	48	49	50
Item 4 required time	86	86	86	86	86	91	91	91	91	91
Item 5 required time	211	211	211	211	211	192	192	192	192	192
Total	297	297	297	297	297	283	283	283	283	283
Item 4 %	0.29					0.32				
Item 5 %	0.71					0.68				

The production requirements for line 3 do not exceed the maximum allowable capacities. In fact, for only five days do they reach the minimum rate of 432 minutes. The master scheduler may therefore decide to run the line at the desired rate until the resource build ahead is calculated to exceed some number of minutes in the next period. Note that any quantity of build-ahead minutes could be selected. In this example, the master scheduler selects 1000 minutes as the maximum build-ahead time. When this maximum is reached, the scheduler will reschedule the line at the minimum rate, or shut it down.

In actual practice, a master scheduler may elect to run a line at a rate that is between the minimum and desired rates, or between the desired and maximum rates.

We have determined the amount of build time that is required. In periods 1–5, it was 432 minutes. We know that we would like to run at the desired rate, which is 480 minutes. The difference (for one day) is 48 minutes. This 48 minutes is build-ahead or hedge time. For 48 minutes today, we are building product that we really do not need today, but the economics of running the line have told us that if we can run the line at the desired rate, we will achieve improved financial results.

The mix percentage of the multiple items being made on the same line can then be applied against the total actual planned rate of 480 minutes.

Since we have added a constraint that the hedge time cannot exceed 1000 minutes in the next period, we must also take the daily hedge time and maintain a cumulative total. When the 1000 maximum is reached, we can shut the line down and consume the hedge time until an insufficient amount of hedge time remains to satisfy the line required time.

Start up of the line (after shut down) could be constrained to the desired rate, the minimum rate, or any interim rate. In this example, we allow start up at the minimum rate.

Table 11.8 illustrates the calculations that have been discussed.
Note that the line has been planned to be shut down four times for a total of nine days. Also remember that the constraint of a hedge time not exceeding 1000 minutes could have been set at 500 minutes, 100 minutes, or even zero.

CONVERTING BUILD TIME TO BUILD QUANTITY

In the previous section, we have calculated the amount of time that we will expend on the production of each item being built on the line. By dividing the planned build time for an item by the amount of time required to build one item, we can determine the number of items that we plan to produce (Table 11.9). This is necessary since some of the end items may have components that are produced in a job shop, lot sized environment.
Table 11.9 illustrates the calculations necessary to convert the build times to build quantities. The exercise is necessary to develop the initial master schedules.

TABLE 11.8 Build-Ahead Calculations

				Periods in Days						
	1	2	3	4	5	6	7	8	9	10
Line 3 required time	432	432	432	432	432	418	418	418	418	418
Item 4 %	0.11					0.13				
Item 5 %	0.89					0.87				
Rate	480	480	480	480	480	480	480	480	480	480
Current build ahead	48	48	48	48	48	62	62	62	62	62
Cumulative build ahead	48	96	144	192	240	302	364	426	488	550
Item 4 build time	53	53	53	53	53	62	62	62	62	62
Item 5 build time	427	427	427	427	427	418	418	418	418	418

				Periods in Days						
	11	12	13	14	15	16	17	18	19	20
Line 3 required time	404	404	404	404	404	388	388	388	388	388
Item 4 %	0.14					0.16				
Item 5 %	0.86					0.84				
Rate	480	480	480	480	480	0	0	432	432	432
Current build ahead	76	76	76	76	76	0	0	44	44	44
Cumulative build ahead	626	702	778	854	930	542	154	198	242	286
Item 4 build time	67	67	67	67	67	0	0	69	69	69
Item 5 build time	413	413	413	413	413	0	0	363	363	363

TABLE 11.8 (Continued)

	Periods in Days									
	21	22	23	24	25	26	27	28	29	30
Line 3 required time	374	374	374	374	374	341	341	341	341	341
Item 4 %	0.17					0.21				
Item 5 %	0.83					0.79				
Rate	432	432	432	432	432	432	432	432	432	0
Current build ahead	58	58	58	58	58	91	91	91	91	0
Cumulative build ahead	344	402	460	518	576	667	758	849	940	599
Item 4 build time	73	73	73	73	73	91	91	91	91	0
Item 5 build time	359	359	359	359	359	341	341	341	341	0

	Periods in Days									
	31	32	33	34	35	36	37	38	39	40
Line 3 required time	327	327	327	327	327	312	312	312	312	312
Item 4 %	0.24					0.26				
Item 5 %	0.76					0.74				
Rate	0	432	432	432	432	432	432	0	0	432
Current build ahead	0	105	105	105	105	120	120	0	0	120
Cumulative build ahead	272	377	482	587	692	812	932	620	308	428
Item 4 build time	0	104	104	104	104	112	112	0	0	112
Item 5 build time	0	328	328	328	328	320	320	0	0	320

Periods in Days

	41	42	43	44	45	46	47	48	49	50
Lines 3 required time	297	297	297	297	297	283	283	283	283	283
Item 4%	0.29					0.32				
Item 5%	0.71					0.68				
Rate	432	432	432	432	0	0	0	432	432	432
Current build ahead	135	135	135	135	0	0	0	149	149	149
Cumulative build ahead	563	698	833	968	671	388	105	254	403	552
Item 4 build time	125	125	125	125	0	0	0	138	138	138
Item 5 build time	307	307	307	307	0	0	0	294	294	294

143

TABLE 11.9 Time to Quantity Conversions

					Periods in Days					
	1	2	3	4	5	6	7	8	9	10
Item 4 build time/by 0.24 per item	53	53	53	53	53	62	62	62	62	62
= Planned quantity	221	221	221	221	221	258	258	258	258	258
Item 5 Build Time/by 0.96 per item	427	427	427	427	427	418	418	418	418	418
= Planned quantity	445	445	445	445	445	435	435	435	435	435

					Periods in Days					
	11	12	13	14	15	16	17	18	19	20
Item 4 build time/by 0.24 per item	67	67	67	67	67	0	0	69	69	69
= Planned quantity	279	279	279	279	279	0	0	288	288	288
Item 5 build time/by 0.96 per item	413	413	413	413	413	0	0	363	363	363
= Planned quantity	430	430	430	430	430	0	0	378	378	378

					Periods in Days					
	21	22	23	24	25	26	27	28	29	30
Item 4 build time/by 0.24 per item	73	73	73	73	73	91	91	91	91	0
= Planned quantity	304	304	304	304	304	379	379	379	379	0
Item 5 build time/by 0.96 per item	359	359	359	359	359	341	341	341	341	0
= Planned quantity	374	374	374	374	374	355	355	355	355	0

Periods in Days

	31	32	33	34	35	36	37	38	39	40
Item 4 build time/by 0.24 per item	0	104	104	104	104	112	112	0	0	112
= Planned quantity	0	433	433	433	433	467	467	0	0	467
Item 5 build time/by 0.96 per item	0	328	328	328	328	320	320	0	0	320
= Planned quantity	0	342	342	342	342	333	333	0	0	333

Periods in Days

	41	42	43	44	45	46	47	48	49	50
Item 4 build time/by 0.24 per item	125	125	125	125	0	0	0	138	138	138
= Planned quantity	521	521	521	521	0	0	0	575	575	575
Item 5 build time/by 0.96 per item	307	307	307	307	0	0	0	294	294	294
= Planned quantity	320	320	320	320	0	0	0	306	306	306

TABLE 11.10   Initial Master Schedules

Item	\multicolumn{10}{c}{Periods in Days}									
	1	2	3	4	5	6	7	8	9	10
1	10					10		10		10
2	40	40	40	40	40	40	40	40	40	40
3	30	30	30	30	30	30	30	30	30	30
4	221	221	221	221	221	258	258	258	258	258
5	445	445	445	445	445	435	435	435	435	435
6	200		200			200	200		200	
7										
8	500			500			500			500
9	400	400	400	400	400	400	600	400	600	400

Item	\multicolumn{10}{c}{Periods in Days}									
	11	12	13	14	15	16	17	18	19	20
1	10	10	10	10	10	20	10	20	10	20
2	40	40	40	40	40	40	40	40	40	40
3	30	30	30	30	30	30	30	30	30	
4	279	279	279	279	279			288	288	288
5	430	430	430	430	430			378	378	378
6	200	200	200		200	200	200		200	200
7	300								300	
8			500			500			500	
9	600	600	600	600	600	800	600	800	600	800

Item	\multicolumn{10}{c}{Periods in Days}									
	21	22	23	24	25	26	27	28	29	30
1	10	20	10	20	10	10	10	20	10	10
2	40	40	40	40	40	40	40	40	40	40
3	30	30	30	30	30	30	30		30	30
4	304	304	304	304	304	379	379	379	379	
5	374	374	374	374	374	355	355	355	355	
6	200	200	200	200		200	200	200	200	200
7						300				
8			500					500		
9	800	800	800	800	800	1000	800	1000	800	1000

TABLE 11.10   (Continued)

	Periods in Days									
Item	31	32	33	34	35	36	37	38	39	40
1	10	10	10	20	10	10	10		10	10
2	40	40	40	40	40	40	40	40	40	40
3	30	30	30	30	30	30	30	60	30	30
4		433	433	433	433	467	467			467
5		342	342	342	342	333	333			333
6	400	200	200	200	200	200	200	400	200	200
7	300				300				300	
8				500						
9	1000	1000	1000	1000	800	1000	1000	800	1000	800

	Periods in Days									
Item	41	42	43	44	45	46	47	48	49	50
1	10		10		10	10			10	
2	40	40	40	40	40	40	40	40	40	40
3	30	30	30	30	30	60	30	30	30	30
4	521	521	521	521				575	575	575
5	320	320	320	320				306	306	306
6	200	400	200	200	200	400	200	400	200	200
7		300				300			300	
8		500								
9	800	800	800	800	800	600	800	600	800	600

THE END ITEM MASTER SCHEDULE PLAN SUMMARY

Table 11.10 is a summary of the master schedule plans that have been developed. It includes all items, whether scheduled by lot sizing criteria or rates. This concludes the chapter on the initial development of a master schedule. In the next chapter we attempt to prove that this schedule can actually be produced. We will perform some rough-cut capacity testing. As the name implies, this testing is still at a very general or rough level.

In previous chapters, we have built plans at a family level (the family production plans) and tested them at a family level with resource requirements testing. We took the tested data and built a plan at an item level (the master schedule) and will now test it with rough-cut capacity testing. We are refining the plan in our example from a very general level to a very finite level. Each level must show that the plan has a reasonable chance of being achieved or the data cannot be passed to the next lower level.

CHAPTER TWELVE

RESOURCE TESTING THE MASTER SCHEDULE

ITEM RESOURCE PROFILES

The item resource profiles in Table 12.1 have been copied from a previous section. Note that this is an interesting chart to have available for the master scheduler. For each end item, it depicts the required resources (in time) that are necessary to produce the item.

You should note that while four of the end items require only work center (job shop) resources, the remainder require a mixture of both job and flow shop resources.

These profiles, which were used as the basis for the family profiles in production planning, will now be used for resource testing the master schedule.

REQUIRED RESOURCES FOR THE MASTER SCHEDULE

For each item, the required production quantities have been developed. These quantities must be extended by the actual planned item production run times per period. In Table 12.2, the amount of time that an end item requires in a facility is extended by the master schedule quantities. This is a very long table, and is not too practical to do without a computer. However, bear with this long table and you will see how a fantastic plan may fall apart when it is resource tested.

Per period, we have converted the planned quantities for each item into the amounts of required time. This has been done by facility. To understand the

TABLE 12.1 Item Resource Profiles

| End Item | W/C 1 | W/C 2 | W/C 3 | W/C 4 | Run Time Requirements | | | | |
					Line 1	Line 2	Line 3	Line 4	Line 5
1	0.10		4.20	6.00					
2	0.10		4.20	6.00					
3	0.10		4.20	6.00					
4	0.10				0.32	0.48	0.24		
5	0.10				0.32	0.48	0.96		
6		1.08			0.32			0.96	0.05
7		1.08			0.32			0.96	0.05
8		0.84			0.32			0.96	0.05
9		0.60							

TABLE 12.2 Resource Testing the Master Schedule

					Period					
Item 1	1	2	3	4	5	6	7	8	9	10
Master schedule quantities	10					10		10		10
× the W/C 1 run time requirement of 0.10 minutes =	1					1		1		1
× the W/C 3 run time requirement of 4.20 minutes =	42					42		42		42
× the W/C 4 run time requirement of 6.00 minutes =	60					60		60		60

					Period					
Item 1	11	12	13	14	15	16	17	18	19	20
Master schedule quantities	10	10	10	10	10	20	10	20	10	20
× the W/C 1 run time requirement of 0.10 minutes =	1	1	1	1	1	2	1	2	1	2
× the W/C 3 run time requirement of 4.20 minutes =	42	42	42	42	42	84	42	84	42	84
× the W/C 4 run time requirement of 6.00 minutes =	60	60	60	60	60	120	60	120	60	120

					Period					
Item 1	21	22	23	24	25	26	27	28	29	30
Master schedule quantities	10	20	10	20	10	10	10	20	10	10
× the W/C 1 run time requirement of 0.10 minutes =	1	2	1	2	1	1	1	2	1	1
× the W/C 3 run time requirement of 4.20 minutes =	42	84	42	84	42	42	42	84	42	42
× the W/C 4 run time requirement of 6.00 minutes =	60	120	60	120	60	60	60	120	60	60

					Period					
Item 1	31	32	33	34	35	36	37	38	39	40
Master schedule quantities	10	10	10	20	10	10	10		10	10
× the W/C 1 run time requirement of 0.10 minutes =	1	1	1	2	1	1	1		1	1
× the W/C 3 run time requirement of 4.20 minutes =	42	42	42	84	42	42	42		42	42
× the W/C 4 run time requirement of 6.00 minutes =	60	60	60	120	60	60	60		60	60

TABLE 12.2   (Continued)

				Period						
Item 1	41	42	43	44	45	46	47	48	49	50

Master schedule quantities

10		10		10	10			10	

× the W/C 1 run time requirement of 0.10 minutes =

1		1		1	1			1	

× the W/C 3 run time requirement of 4.20 minutes =

42		42		42	42			42	

× the W/C 4 run time requirement of 6.00 minutes =

60		60		60	60			60	

				Period						
Item 2	1	2	3	4	5	6	7	8	9	10

Master schedule quantities

40	40	40	40	40	40	40	40	40	40

× the W/C 1 run time requirement of 0.10 minutes =

4	4	4	4	4	4	4	4	4	4

× the W/C 3 run time requirement of 4.20 minutes =

168	168	168	168	168	168	168	168	168	168

× the W/C 4 run time requirement of 6.00 minutes =

240	240	240	240	240	240	240	240	240	240

				Period						
Item 2	11	12	13	14	15	16	17	18	19	20

Master schedule quantities

40	40	40	40	40	40	40	40	40	40

× the W/C 1 run time requirement of 0.10 minutes =

4	4	4	4	4	4	4	4	4	4

× the W/C 3 run time requirement of 4.20 minutes =

168	168	168	168	168	168	168	168	168	168

× the W/C 4 run time requirement of 6.00 minutes =

240	240	240	240	240	240	240	240	240	240

				Period						
Item 2	21	22	23	24	25	26	27	28	29	30

Master schedule quantities

40	40	40	40	40	40	40	40	40	40

× the W/C 1 run time requirement of 0.10 minutes =

4	4	4	4	4	4	4	4	4	4

× the W/C 3 run time requirement of 4.20 minutes =

168	168	168	168	168	168	168	168	168	168

× the W/C 4 run time requirement of 6.00 minutes =

240	240	240	240	240	240	240	240	240	240

TABLE 12.2 (Continued)

						Period				
Item 2	31	32	33	34	35	36	37	38	39	40

Master schedule quantities
| | 40 | 40 | 40 | 40 | 40 | 40 | 40 | 40 | 40 | 40 |

× the W/C 1 run time requirement of 0.10 minutes =
| | 4 | 4 | 4 | 4 | 4 | 4 | 4 | 4 | 4 | 4 |

× the W/C 3 run time requirement of 4.20 minutes =
| | 168 | 168 | 168 | 168 | 168 | 168 | 168 | 168 | 168 | 168 |

× the W/C 4 run time requirement of 6.00 minutes =
| | 240 | 240 | 240 | 240 | 240 | 240 | 240 | 240 | 240 | 240 |

						Period				
Item 2	41	42	43	44	45	46	47	48	49	50

Master schedule quantities
| | 40 | 40 | 40 | 40 | 40 | 40 | 40 | 40 | 40 | 40 |

× the W/C 1 run time requirement of 0.10 minutes =
| | 4 | 4 | 4 | 4 | 4 | 4 | 4 | 4 | 4 | 4 |

× the W/C 3 run time requirement of 4.20 minutes =
| | 168 | 168 | 168 | 168 | 168 | 168 | 168 | 168 | 168 | 168 |

× the W/C 4 run time requirement of 6.00 minutes =
| | 240 | 240 | 240 | 240 | 240 | 240 | 240 | 240 | 240 | 240 |

						Period				
Item 3	1	2	3	4	5	6	7	8	9	10

Master schedule quantities
| | 30 | 30 | 30 | 30 | 30 | 30 | 30 | 30 | 30 | 30 |

× the W/C 1 run time requirement of 0.10 minutes =
| | 3 | 3 | 3 | 3 | 3 | 3 | 3 | 3 | 3 | 3 |

× the W/C 3 run time requirement of 4.20 minutes =
| | 126 | 126 | 126 | 126 | 126 | 126 | 126 | 126 | 126 | 126 |

× the W/C 4 run time requirement of 6.00 minutes =
| | 180 | 180 | 180 | 180 | 180 | 180 | 180 | 180 | 180 | 180 |

						Period				
Item 3	11	12	13	14	15	16	17	18	19	20

Master schedule quantities
| | 30 | 30 | 30 | 30 | 30 | 30 | 30 | 30 | 30 | |

× the W/C 1 run time requirement of 0.10 minutes =
| | 3 | 3 | 3 | 3 | 3 | 3 | 3 | 3 | 3 | |

× the W/C 3 run time requirement of 4.20 minutes =
| | 126 | 126 | 126 | 126 | 126 | 126 | 126 | 126 | 126 | |

× the W/C 4 run time requirement of 6.00 minutes =
| | 180 | 180 | 180 | 180 | 180 | 180 | 180 | 180 | 180 | |

TABLE 12.2   (Continued)

						Period				
Item 3	21	22	23	24	25	26	27	28	29	30
Master schedule quantities	30	30	30	30	30	30	30		30	30
× the W/C 1 run time requirement of 0.10 minutes =	3	3	3	3	3	3	3		3	3
× the W/C 3 run time requirement of 4.20 minutes =	126	126	126	126	126	126	126		126	126
× the W/C 4 run time requirement of 6.00 minutes =	180	180	180	180	180	180	180		180	180

						Period				
Item 3	31	32	33	34	35	36	37	38	39	40
Master schedule quantities	30	30	30	30	30	30	30	60	30	30
× the W/C 1 run time requirement of 0.10 minutes =	3	3	3	3	3	3	3	6	3	3
× the W/C 3 run time requirement of 4.20 minutes =	126	126	126	126	126	126	126	252	126	126
× the W/C 4 run time requirement of 6.00 minutes =	180	180	180	180	180	180	180	360	180	180

						Period				
Item 3	41	42	43	44	45	46	47	48	49	50
Master schedule quantities	30	30	30	30	30	60	30	30	30	30
× the W/C 1 run time requirement of 0.10 minutes =	3	3	3	3	3	6	3	3	3	3
× the W/C 3 run time requirement of 4.20 minutes =	126	126	126	126	126	252	126	126	126	126
× the W/C 4 run time requirement of 6.00 minutes =	180	180	180	180	180	360	180	180	180	180

						Period				
Item 4	1	2	3	4	5	6	7	8	9	10
Master schedule quantities	221	221	221	221	221	258	258	258	258	258
× the W/C 1 run time requirement of 0.10 minutes =	22	22	22	22	22	26	26	26	26	26
× the line 1 run time requirement of 0.32 minutes =	71	71	71	71	71	83	83	83	83	83
× the line 2 run time requirement of 0.48 minutes =	106	106	106	106	106	124	124	124	124	124
× the line 3 run time requirement of 0.24 minutes =	53	53	53	53	53	62	62	62	62	62

TABLE 12.2   (Continued)

Item 4	Period									
	11	12	13	14	15	16	17	18	19	20
Master schedule quantities										
	279	279	279	279	279			288	288	288
× the W/C 1 run time requirement of 0.10 minutes =										
	28	28	28	28	28			29	29	29
× the line 1 run time requirement of 0.32 minutes =										
	89	89	89	89	89			92	92	92
× the line 2 run time requirement of 0.48 minutes =										
	134	134	134	134	134			138	138	138
× the line 3 run time requirement of 0.24 minutes =										
	67	67	67	67	67			69	69	69

Item 4	Period									
	21	22	23	24	25	26	27	28	29	30
Master schedule quantities										
	304	304	304	304	304	379	379	379	379	
× the W/C 1 run time requirement of 0.10 minutes =										
	30	30	30	30	30	38	38	38	38	
× the line 1 run time requirement of 0.32 minutes =										
	97	97	97	97	97	121	121	121	121	
× the line 2 run time requirement of 0.48 minutes =										
	146	146	146	146	146	182	182	182	182	
× the line 3 run time requirement of 0.24 minutes =										
	73	73	73	73	73	91	91	91	91	

Item 4	Period									
	31	32	33	34	35	36	37	38	39	40
Master schedule quantities										
		433	433	433	433	467	467			467
× the W/C 1 run time requirement of 0.10 minutes =										
		43	43	43	43	47	47			47
× the line 1 run time requirement of 0.32 minutes =										
		139	139	139	139	149	149			149
× the line 2 run time requirement of 0.48 minutes =										
		208	208	208	208	224	224			224
× the line 3 run time requirement of 0.24 minutes =										
		104	104	104	104	112	112			112

TABLE 12.2 (Continued)

				Period					
Item 4	41	42	43	44	45	46	47	48	49	50

Master schedule quantities
| | 521 | 521 | 521 | 521 | | | | 575 | 575 | 575 |

× the W/C 1 run time requirement of 0.10 minutes =
| | 52 | 52 | 52 | 52 | | | | 58 | 58 | 58 |

× the line 1 run time requirement of 0.32 minutes =
| | 167 | 167 | 167 | 167 | | | | 184 | 184 | 184 |

× the line 2 run time requirement of 0.48 minutes =
| | 250 | 250 | 250 | 250 | | | | 276 | 276 | 276 |

× the line 3 run time requirement of 0.24 minutes =
| | 125 | 125 | 125 | 125 | | | | 138 | 138 | 138 |

				Period					
Item 5	1	2	3	4	5	6	7	8	9	10

Master schedule quantities
| | 445 | 445 | 445 | 445 | 445 | 435 | 435 | 435 | 435 | 435 |

× the W/C 1 run time requirement of 0.10 minutes =
| | 45 | 45 | 45 | 45 | 45 | 44 | 44 | 44 | 44 | 44 |

× the line 1 run time requirement of 0.32 minutes =
| | 142 | 142 | 142 | 142 | 142 | 139 | 139 | 139 | 139 | 139 |

× the line 2 run time requirement of 0.48 minutes =
| | 214 | 214 | 214 | 214 | 214 | 209 | 209 | 209 | 209 | 209 |

× the line 3 run time requirement of 0.96 minutes =
| | 427 | 427 | 427 | 427 | 427 | 418 | 418 | 418 | 418 | 418 |

				Period					
Item 5	11	12	13	14	15	16	17	18	19	20

Master schedule quantities
| | 430 | 430 | 430 | 430 | 430 | | | 378 | 378 | 378 |

× the W/C 1 run time requirement of 0.10 minutes =
| | 43 | 43 | 43 | 43 | 43 | | | 38 | 38 | 38 |

× the line 1 run time requirement of 0.32 minutes =
| | 138 | 138 | 138 | 138 | 138 | | | 121 | 121 | 121 |

× the line 2 run time requirement of 0.48 minutes =
| | 206 | 206 | 206 | 206 | 206 | | | 181 | 181 | 181 |

× the line 3 run time requirement of 0.96 minutes =
| | 413 | 413 | 413 | 413 | 413 | | | 363 | 363 | 363 |

TABLE 12.2 (Continued)

				Period					
Item 5	21	22	23	24	25	26	27	28	29	30

Master schedule quantities

| | 374 | 374 | 374 | 374 | 374 | 355 | 355 | 355 | 355 | |

× the W/C 1 run time requirement of 0.10 minutes =

| | 37 | 37 | 37 | 37 | 37 | 36 | 36 | 36 | 36 | |

× the line 1 run time requirement of 0.32 minutes =

| | 120 | 120 | 120 | 120 | 120 | 114 | 114 | 114 | 114 | |

× the line 2 run time requirement of 0.48 minutes =

| | 180 | 180 | 180 | 180 | 180 | 170 | 170 | 170 | 170 | |

× the line 3 run time requirement of 0.96 minutes =

| | 359 | 359 | 359 | 359 | 359 | 341 | 341 | 341 | 341 | |

				Period					
Item 5	31	32	33	34	35	36	37	38	39	40

Master schedule quantities

| | | 342 | 342 | 342 | 342 | 333 | 333 | | | 333 |

× the W/C 1 run time requirement of 0.10 minutes =

| | | 34 | 34 | 34 | 34 | 33 | 33 | | | 33 |

× the line 1 run time requirement of 0.32 minutes =

| | | 109 | 109 | 109 | 109 | 107 | 107 | | | 107 |

× the line 2 run time requirement of 0.48 minutes =

| | | 164 | 164 | 164 | 164 | 160 | 160 | | | 160 |

× the line 3 run time requirement of 0.96 minutes =

| | | 328 | 328 | 328 | 328 | 320 | 320 | | | 320 |

				Period					
Item 5	41	42	43	44	45	46	47	48	49	50

Master schedule quantities

| | 320 | 320 | 320 | 320 | | | | 306 | 306 | 306 |

× the W/C 1 run time requirement of 0.10 minutes =

| | 32 | 32 | 32 | 32 | | | | 31 | 31 | 31 |

× the line 1 run time requirement of 0.32 minutes =

| | 102 | 102 | 102 | 102 | | | | 98 | 98 | 98 |

× the line 2 run time requirement of 0.48 minutes =

| | 154 | 154 | 154 | 154 | | | | 147 | 147 | 147 |

× the line 3 run time requirement of 0.96 minutes =

| | 307 | 307 | 307 | 307 | | | | 294 | 294 | 294 |

TABLE 12.2 (Continued)

					Period					
Item 6	1	2	3	4	5	6	7	8	9	10
Master schedule quantities	200		200			200	200		200	
× the W/C 2 run time requirement of 1.08 minutes =	216		216			216	216		216	
× the line 1 run time requirement of 0.32 minutes =	64		64			64	64		64	
× the line 4 run time requirement of 0.96 minutes =	192		192			192	192		192	
× the line 5 run time requirement of 0.05 minutes =	10		10			10	10		10	

					Period					
Item 6	11	12	13	14	15	16	17	18	19	20
Master schedule quantities	200	200	200		200	200	200		200	200
× the W/C 2 run time requirement of 1.08 minutes =	216	216	216		216	216	216		216	216
× the line 1 run time requirement of 0.32 minutes =	64	64	64		64	64	64		64	64
× the line 4 run time requirement of 0.96 minutes =	192	192	192		192	192	192		192	192
× the line 5 run time requirement of 0.05 minutes =	10	10	10		10	10	10		10	10

					Period					
Item 6	21	22	23	24	25	26	27	28	29	30
Master schedule quantities	200	200	200	200		200	200	200	200	200
× the W/C 2 run time requirement of 1.08 minutes =	216	216	216	216		216	216	216	216	216
× the line 1 run time requirement of 0.32 minutes =	64	64	64	64		64	64	64	64	64
× the line 4 run time requirement of 0.96 minutes =	192	192	192	192		192	192	192	192	192
× the line 5 run time requirement of 0.05 minutes =	10	10	10	10		10	10	10	10	10

TABLE 12.2   (Continued)

					Period					
Item 6	31	32	33	34	35	36	37	38	39	40

Master schedule quantities
	400	200	200	200	200	200	200	400	200	200

× the W/C 2 run time requirement of 1.08 minutes =
	432	216	216	216	216	216	216	432	216	216

× the line 1 run time requirement of 0.32 minutes =
	128	64	64	64	64	64	64	128	64	64

× the line 4 run time requirement of 0.96 minutes =
	384	192	192	192	192	192	192	384	192	192

× the line 5 run time requirement of 0.05 minutes =
	20	10	10	10	10	10	10	20	10	10

					Period					
Item 6	41	42	43	44	45	46	47	48	49	50

Master schedule quantities
	200	400	200	200	200	400	200	400	200	200

× the W/C 2 run time requirement of 1.08 minutes =
	216	432	216	216	216	432	216	432	216	216

× the line 1 run time requirement of 0.32 minutes =
	64	128	64	64	64	128	64	128	64	64

× the line 4 run time requirement of 0.96 minutes =
	192	384	192	192	192	384	192	384	192	192

× the line 5 run time requirement of 0.05 minutes =
	10	20	10	10	10	20	10	20	10	10

					Period					
Item 7	1	2	3	4	5	6	7	8	9	10

Master schedule quantities

× the W/C 2 run time requirement of 1.08 minutes =

× the line 1 run time requirement of 0.32 minutes =

× the line 4 run time requirement of 0.96 minutes =

× the line 5 run time requirement of 0.05 minutes =

TABLE 12.2   (Continued)

	Period									
Item 7	11	12	13	14	15	16	17	18	19	20

Master schedule quantities
    300   300
× the W/C 2 run time requirement of 1.08 minutes =
    324   324
× the line 1 run time requirement of 0.32 minutes =
    96   96
× the line 4 run time requirement of 0.96 minutes =
    288   288
× the line 5 run time requirement of 0.05 minutes =
    15   15

	Period									
Item 7	21	22	23	24	25	26	27	28	29	30

Master schedule quantities
                                          300
× the W/C 2 run time requirement of 1.08 minutes =
                                          324
× the line 1 run time requirement of 0.32 minutes =
                                          96
× the line 4 run time requirement of 0.96 minutes =
                                          288
× the line 5 run time requirement of 0.05 minutes =
                                          15

	Period									
Item 7	31	32	33	34	35	36	37	38	39	40

Master schedule quantities
    300   300   300
× the W/C 2 run time requirement of 1.08 minutes =
    324   324   324
× the line 1 run time requirement of 0.32 minutes =
    96   96   96
× the line 4 run time requirement of 0.96 minutes =
    288   288   288
× the line 5 run time requirement of 0.05 minutes =
    15   15   15

TABLE 12.2   (Continued)

					Period					
Item 7	41	42	43	44	45	46	47	48	49	50

Master schedule quantities

		300				300			300	

× the W/C 2 run time requirement of 1.08 minutes =

		324				324			324	

× the line 1 run time requirement of 0.32 minutes =

		96				96			96	

× the line 4 run time requirement of 0.96 minutes =

		288				288			288	

× the line 5 run time requirement of 0.05 minutes =

		15				15			15	

					Period					
Item 8	1	2	3	4	5	6	7	8	9	10

Master schedule quantities

	500			500			500			500

× the W/C 2 run time requirement of 0.84 minutes =

	420			420			420			420

× the line 1 run time requirement of 0.32 minutes =

	160			160			160			160

× the line 4 run time requirement of 0.96 minutes =

	480			480			480			480

× the line 5 run time requirement of 0.05 minutes =

	25			25			25			25

					Period					
Item 8	11	12	13	14	15	16	17	18	19	20

Master schedule quantities

		500				500			500	

× the W/C 2 run time requirement of 0.84 minutes =

		420				420			420	

× the line 1 run time requirement of 0.32 minutes =

		160				160			160	

× the line 4 run time requirement of 0.96 minutes =

		480				480			480	

× the line 5 run time requirement of 0.05 minutes =

		25				25			25	

TABLE 12.2 (Continued)

				Period						
Item 8	21	22	23	24	25	26	27	28	29	30

Master schedule quantities
| | | | 500 | | | | | 500 | | |
× the W/C 2 run time requirement of 0.84 minutes =
| | | | 420 | | | | | 420 | | |
× the line 1 run time requirement of 0.32 minutes =
| | | | 160 | | | | | 160 | | |
× the line 4 run time requirement of 0.96 minutes =
| | | | 480 | | | | | 480 | | |
× the line 5 run time requirement of 0.05 minutes =
| | | | 25 | | | | | 25 | | |

				Period						
Item 8	31	32	33	34	35	36	37	38	39	40

Master schedule quantities
| | | | 500 | | | | | | | |
× the W/C 2 run time requirement of 0.84 minutes =
| | | | 420 | | | | | | | |
× the line 1 run time requirement of 0.32 minutes =
| | | | 160 | | | | | | | |
× the line 4 run time requirement of 0.96 minutes =
| | | | 480 | | | | | | | |
× the line 5 run time requirement of 0.05 minutes =
| | | | 25 | | | | | | | |

				Period						
Item 8	41	42	43	44	45	46	47	48	49	50

Master schedule quantities
| | 500 | | | | | | | | |
× the W/C 2 run time requirement of 0.84 minutes =
| | 420 | | | | | | | | |
× the line 1 run time requirement of 0.32 minutes =
| | 160 | | | | | | | | |
× the line 4 run time requirement of 0.96 minutes =
| | 480 | | | | | | | | |
× the line 5 run time requirement of 0.05 minutes =
| | 25 | | | | | | | | |

TABLE 12.2   **(Continued)**

					Period				
Item 9	1	2	3	4	5	6	7	8	9	10

Master schedule quantities

400	400	400	400	400	400	600	400	600	400

× the W/C 2 run time requirement of 0.60 minutes =

240	240	240	240	240	240	360	240	360	240

					Period				
Item 9	11	12	13	14	15	16	17	18	19	20

Master schedule quantities

600	600	600	600	600	800	600	800	600	800

× the W/C 2 run time requirement of 0.60 minutes =

360	360	360	360	360	480	360	480	360	480

					Period				
Item 9	21	22	23	24	25	26	27	28	29	30

Master schedule quantities

800	800	800	800	800	1000	800	1000	800	1000

× the W/C 2 run time requirement of 0.60 minutes =

480	480	480	480	480	600	480	600	480	600

					Period				
Item 9	31	32	33	34	35	36	37	38	39	40

Master schedule quantities

1000	1000	1000	1000	800	1000	1000	800	1000	800

× the W/C 2 run time requirement of 0.60 minutes =

600	600	600	600	480	600	600	480	600	480

					Period				
Item 9	41	42	43	44	45	46	47	48	49	50

Master schedule quantities

800	800	800	800	800	600	800	600	800	600

× the W/C 2 run time requirement of 0.60 minutes =

480	480	480	480	480	360	480	360	480	360

impact on a facility, it will be necessary to regroup this data by item within facility.

REQUIRED ITEM RESOURCES SUMMARY (ROUGH-CUT CAPACITY TEST)

The resources required by each facility are accumulated and shown in Table 12.3.

These required resources can be equated to the available resources (in minutes) and overages can be shown where they occur. An overage will require (in the future) an adjustment to one or more of the item production plans.

As overages are indicated for a specific facility, the reader should reflect on how, if he or she were the master scheduler, the previously planned item quantities should be adjusted (to eliminate the overages) without creating overages in other facilities that were previously within resource availability limitations.

It is obvious that if one reviews the overages in the previous plan, they may easily conclude that it is invalid, and should not be passed to material requirements planning.

What appeared to be a good plan in previous sections now seems to fall apart. This is not all wrong and in the next chapters, we will discuss what should be done to correct the situation. (It really is a multi-iteration process.)

In addition, we should consider some alternative approaches that might be applied.

And, of course, at this time we do not have a valid plan to produce product for our example. In this text, we will not develop that complete example. We have built a very simply case study production example, and even with the simplicity, the tables became lengthy. Many more iterations remain to be performed before master schedule planning can be considered to be complete, but since the logic will remain the same as already illustrated, it would have minimum value to develop additional tables for the trial fitting of different production plan approaches.

Some suggestions will be provided for the reader to complete the example to the shop floor level, as well as some ideas for extending the example.

TABLE 12.3   Summary of Facility Required Resources

Facility: W/C 1   *Available Resource per Period: 480*

					Period					
Item	1	2	3	4	5	6	7	8	9	10
1	1					1		1		1
2	4	4	4	4	4	4	4	4	4	4
3	3	3	3	3	3	3	3	3	3	3
4	22	22	22	22	22	26	26	26	26	26
5	45	45	45	45	45	44	44	44	44	44
Total	75	74	74	74	74	78	77	78	77	78

					Period					
Item	11	12	13	14	15	16	17	18	19	20
1	1	1	1	1	1	2	1	2	1	2
2	4	4	4	4	4	4	4	4	4	4
3	3	3	3	3	3	3	3	3	3	3
4	28	28	28	28	28			29	29	29
5	43	43	43	43	43			38	38	38
Total	79	79	79	79	79	9	8	76	75	76

					Period					
Item	21	22	23	24	25	26	27	28	29	30
1	1	2	1	2	1	1	1	2	1	1
2	4	4	4	4	4	4	4	4	4	4
3	3	3	3	3	3	3	3		3	3
4	30	30	30	30	30	38	38	38	38	
5	37	37	37	37	37	36	36	36	36	
Total	75	76	75	76	75	82	82	80	82	8

					Period					
Item	31	32	33	34	35	36	37	38	39	40
1	1	1	1	2	1	1	1		1	1
2	4	4	4	4	4	4	4	4	4	4
3	3	3	3	3	3	3	3	6	3	3
4		43	43	43	43	47	47			47
5		34	34	34	34	33	33			33
Total	8	85	85	86	85	88	88	10	8	88

TABLE 12.3 (Continued)

	Period									
Item	41	42	43	44	45	46	47	48	49	50
1	1		1		1	1			1	
2	4	4	4	4	4	4	4	4	4	4
3	3	3	3	3	3	6	3	3	3	3
4	52	52	52	52				58	58	58
5	32	32	32	32				31	31	31
Total	92	91	92	91	8	11	7	96	97	96

Facility: W/C 2   *Available Resource per Period: 960*

	Period									
Item	1	2	3	4	5	6	7	8	9	10
6	216		216			216	216		216	
7										
8	420			420			420			420
9	240	240	240	240	240	240	360	240	360	240
Total	876	240	456	660	240	456	996	240	576	660
Overage							36			

	Period									
Item	11	12	13	14	15	16	17	18	19	20
6	216	216	216		216	216	216		216	216
7	324								324	
8			420			420			420	
9	360	360	360	360	360	480	360	480	360	480
Total	900	576	996	360	576	1116	576	480	1320	696
Overage			36			156			360	

	Period									
Item	21	22	23	24	25	26	27	28	29	30
6	216	216	216	216		216	216	216	216	216
7						324				
8			420					420		
9	480	480	480	480	480	600	480	600	480	600
Total	696	696	1116	696	480	1140	696	1236	696	816
Overage			156			180		276		

TABLE 12.3 (Continued)

| | Period | | | | | | | | | |
Item	31	32	33	34	35	36	37	38	39	40
6	432	216	216	216	216	216	216	432	216	216
7	324				324				324	
8				420						
9	600	600	600	600	480	600	600	480	600	480
Total	1356	816	816	1236	1020	816	816	912	1140	696
Overage	396			276	60				180	

| | Period | | | | | | | | | |
Item	41	42	43	44	45	46	47	48	49	50
6	216	432	216	216	216	432	216	432	216	216
7		324				324			324	
8		420								
9	480	480	480	480	480	360	480	360	480	360
Total	696	1656	696	696	696	1116	696	792	1020	576
Overage		696				156			60	

Facility: W/C 3   *Available Resource per Period: 480*

| | Period | | | | | | | | | |
Item	1	2	3	4	5	6	7	8	9	10
1	42					42		42		42
2	168	168	168	168	168	168	168	168	168	168
3	126	126	126	126	126	126	126	126	126	126
Total	336	294	294	294	294	336	294	336	294	336

| | Period | | | | | | | | | |
Item	11	12	13	14	15	16	17	18	19	20
1	42	42	42	42	42	84	42	84	42	84
2	168	168	168	168	168	168	168	168	168	168
3	126	126	126	126	126	126	126	126	126	
Total	336	336	336	336	336	378	336	378	336	252

| | Period | | | | | | | | | |
Item	21	22	23	24	25	26	27	28	29	30
1	42	84	42	84	42	42	42	84	42	42
2	168	168	168	168	168	168	168	168	168	168
3	126	126	126	126	126	126	126		126	126
Total	336	378	336	378	336	336	336	252	336	336

TABLE 12.3   (Continued)

					Period					
Item	31	32	33	34	35	36	37	38	39	40
1	42	42	42	84	42	42	42		42	42
2	168	168	168	168	168	168	168	168	168	168
3	126	126	126	126	126	252	126	252	126	126
Total	336	336	336	378	336	336	336	420	336	336

					Period					
Item	41	42	43	44	45	46	47	48	49	50
1	42		42		42	42			42	
2	168	168	168	168	168	168	168	168	168	168
3	126	126	126	126	126	252	126	126	126	126
Total	336	294	336	294	336	462	294	294	336	294

Facility: W/C 4   *Available Resource per Period: 480*

					Period					
Item	1	2	3	4	5	6	7	8	9	10
1	60					60		60		60
2	240	240	240	240	240	240	240	240	240	240
3	180	180	180	180	180	180	180	180	180	180
Total	480	420	420	420	420	480	420	480	420	480

					Period					
Item	11	12	13	14	15	16	17	18	19	20
1	60	60	60	60	60	120	60	120	60	120
2	240	240	240	240	240	240	240	240	240	240
3	180	180	180	180	180	180	180	180	180	
Total	480	480	480	480	480	540	480	540	480	360
Overage						60		60		

					Period					
Item	21	22	23	24	25	26	27	28	29	30
1	60	120	60	120	60	60	60	120	60	60
2	240	240	240	240	240	240	240	240	240	240
3	180	180	180	180	180	180		180	180	180
Total	480	540	480	540	480	480	480	360	480	480
Overage		60		60						

TABLE 12.3   (Continued)

					Period					
Item	31	32	33	34	35	36	37	38	39	40
1	60	60	60	120	60	60	60		60	60
2	240	240	240	240	240	240	240	240	240	240
3	180	180	180	180	180	180	180	360	180	180
Total	480	480	480	540	480	480	480	600	480	480
Overage				60				120		

					Period					
Item	41	42	43	44	45	46	47	48	49	50
1	60		60		60	60			60	
2	240	240	240	240	240	240	240	240	240	240
3	180	180	180	180	180	360	180	180	180	180
Total	480	420	480	420	480	660	420	420	480	420
Overage						180				

Facility: Line 1   *Available Resource per Period: 480*

					Period					
Item	1	2	3	4	5	6	7	8	9	10
4	71	71	71	71	71	83	83	83	83	83
5	142	142	142	142	142	139	139	139	139	139
6	64		64			64	64		64	
7										
8	160			160			160			160
Total	437	213	277	373	213	286	446	222	286	382

					Period					
Item	11	12	13	14	15	16	17	18	19	20
4	89	89	89	89	89			92	92	92
5	138	138	138	138	138			121	121	121
6	64	64	64		64	64	64		64	64
7	96								96	
8			160			160			160	
Total	387	291	451	227	291	224	64	213	533	277
Overage									53	

TABLE 12.3 (Continued)

	Period									
Item	21	22	23	24	25	26	27	28	29	30
4	97	97	97	97	97	121	121	121	121	
5	120	120	120	120	120	114	114	114	114	
6	64	64	64	64		64	64	64	64	64
7						96				
8			160						160	
Total	281	281	441	281	217	395	299	459	299	64

	Period									
Item	31	32	33	34	35	36	37	38	39	40
4		139	139	139	139	149	149			149
5		109	109	109	109	107	107			107
6	128	64	64	64	64	64	64	128	64	64
7	96				96				96	
8				160						
Total	224	312	312	472	408	320	320	128	160	320

	Period									
Item	41	42	43	44	45	46	47	48	49	50
4	167	167	167	167				184	184	184
5	102	102	102	102				98	98	98
6	64	128	64	64	64	128	64	128	64	64
7		96				96			96	
8		160								
Total	333	653	333	333	64	224	64	410	442	346
Overage		173								

Facility: Line 2   *Available Resource per Period: 480*

	Period									
Item	1	2	3	4	5	6	7	8	9	10
4	106	106	106	106	106	124	124	124	124	124
5	214	214	214	214	214	209	209	209	209	209
Total	320	320	320	320	320	333	333	333	333	333

	Period									
Item	11	12	13	14	15	16	17	18	19	20
4	134	134	134	134	134			138	138	138
5	206	206	206	206	206			181	181	181
Total	340	340	340	340	340			319	319	319

TABLE 12.3 (Continued)

Item	Period									
	21	22	23	24	25	26	27	28	29	30
4	146	146	146	146	146	182	182	182	182	
5	180	180	180	180	180	170	170	170	170	
Total	326	326	326	326	326	352	352	352	352	

Item	Period									
	31	32	33	34	35	36	37	38	39	40
4		208	208	208	208	224	224			224
5		164	164	164	164	160	160			160
Total		372	372	372	372	384	384			384

Item	Period									
	41	42	43	44	45	46	47	48	49	50
4	250	250	250	250				276	276	276
5	154	154	154	154				147	147	147
Total	404	404	404	404				423	423	423

Facility: Line 3   *Available Resource per Period: 480*

Item	Period									
	1	2	3	4	5	6	7	8	9	10
4	53	53	53	53	53	62	62	62	62	62
5	427	427	427	427	427	418	418	418	418	418
Total	480	480	480	480	480	480	480	480	480	480

Item	Period									
	11	12	13	14	15	16	17	18	19	20
4	67	67	67	67	67			69	69	69
5	413	413	413	413	413			363	363	363
Total	480	480	480	480	480			432	432	432

Item	Period									
	21	22	23	24	25	26	27	28	29	30
4	73	73	73	73	73	91	91	91	91	
5	359	359	359	359	359	341	341	341	341	
Total	432	432	432	432	432	432	432	432	432	

TABLE 12.3 (Continued)

Item	Period									
	31	32	33	34	35	36	37	38	39	40
4		104	104	104	104	112	112			112
5		328	328	328	328	320	320			320
Total		432	432	432	432	432	432			432

Item	Period									
	41	42	43	44	45	46	47	48	49	50
4	125	125	125	125				138	138	138
5	307	307	307	307				294	294	294
Total	432	432	432	432				432	432	432

Facility: Line 4   *Available Resource per Period: 480*

Item	Period									
	1	2	3	4	5	6	7	8	9	10
6	192		192			192	192		192	
7										
8	480			480			480			480
Total	672		192	480		192	672		192	480
Overage	192						192			

Item	Period									
	11	12	13	14	15	16	17	18	19	20
6	192	192	192		192	192	192		192	192
7	288								288	
8			480			480			480	
Total	480	192	672		192	672	192		960	192
Overage			192			192			480	

Item	Period									
	21	22	23	24	25	26	27	28	29	30
6	192	192	192	192		192	192	192	192	192
7						288				
8			480					480		
Total	192	192	672	192		480	192	672	192	192
Overage			192					192		

TABLE 12.3 (Continued)

	Period									
Item	31	32	33	34	35	36	37	38	39	40
6	384	192	192	192	192	192	192	384	192	192
7	288				288				288	
8				480						
Total	672	192	192	672	480	192	192	384	480	192
Overage	**192**			**192**						

	Period									
Item	41	42	43	44	45	46	47	48	49	50
6	192	384	192	192	192	384	192	384	192	192
7		288				288			288	
8		480								
Total	192	1152	192	192	192	672	192	384	480	192
Overage		**672**				**192**				

Facility: Line 5   *Available Resource per Period: 480*

	Period									
Item	1	2	3	4	5	6	7	8	9	10
6	10		10			10	10		10	
7										
8	25			25			25			25
Total	35		10	25		10	35		10	25

	Period									
Item	11	12	13	14	15	16	17	18	19	20
6	10	10	10		10	10	10		10	10
7	15								15	
8			25			25			25	
Total	25	10	35		10	35	10		50	10

	Period									
Item	21	22	23	24	25	26	27	28	29	30
6	10	10	10	10		10	10	10	10	10
7						15				
8			25					25		
Total	10	10	35	10		25	10	35	10	10

TABLE 12.3   (Continued)

Item	Period									
	31	32	33	34	35	36	37	38	39	40
6	20	10	10	10	10	10	10	20	10	10
7	15				15				15	
8				25						
Total	35	10	10	35	25	10	10	20	25	10

Item	Period									
	41	42	43	44	45	46	47	48	49	50
6	10	20	10	10	10	20	10	20	10	10
7		15				15			15	
8		25								
Total	10	60	10	10	10	35	10	20	25	10

CHAPTER THIRTEEN

THE CASE STUDY EXAMPLE—CONCLUSIONS AND RECOMMENDATIONS

CONCLUSIONS FROM THE CASE STUDY

The planning logic for a pure job shop is based on when lot sized order quantities are to be released and received. These quantities are resource tested by the amount of time they consume at work centers. When an overage occurs at a facility, the order quantity can be moved to an earlier start date where unused resource time is available.

In a pure flow shop, where only one item is produced on a particular line, the scheduling is very simple. The line is run between the minimum and maximum rates. When demand drops, the line is shut down. When demand exceeds the maximum rate, either built-aheads are utilized (during a period when demand was low) or another line is added to produce the product.

The schedule planning for this pure flow shop environment is often performed without any computerized scheduling system when only a few end items are involved. When many items are involved, family planning may be performed at the production planning level to develop item production plans, but no master schedule planning is done. The item production plans are sufficient to drive the plant.

The case study example has illustrated an environment where:

Some end items are totally produced in a job shop
Some end items are totally produced in a flow shop

Some end items are produced using a combination of job and flow shop facilities

None of the above conditions has really caused any deviation in basic high level planning logic. The real need for a change in logic happens when multiple items with different build times are planned for the same line. Now you cannot simply plan quantities, but must convert the quantities to time while you are performing the planning function.

In the "Summary of Facility Required Resources" in the previous chapter, look at line 4. There is plenty of room to do build-aheads. For example, period 13 has an overage that could easily be done in period 12. In actuality, the master scheduler would start at the beginning of this schedule and pull forward to fill daily run rates between the minimum and maximum rates. This means that when the line would be scheduled to run, it would run as close to the consumption of 480 minutes as possible. It would also mean that the line would be shut down on many more days than currently shown on the table. And, of course, the amount of build-ahead inventory would increase.

Work center 4 has a more severe problem. Period 46 has an overage of 180, of which 120 can be absorbed by periods 44 and 42. The remaining 60 is added to 120 from period 38 and 60 from period 34 providing a total of 240. Of this amount, 120 can be absorbed by period 28, but then another 120 is picked up in period 24 and 22. By following this logic, you can see that build-aheads must start in period 5. Many companies would consider this to be a plan for the development of excessive hedge inventory, and if they could not subcontract the work closer to the point of need, they would attempt to work with their customers to reschedule the customer orders.

BALANCING THE PLANT

The ideal situation is to totally balance the plant to produce a desired amount of products. This can easily be done today with the many simulation and modeling tools on the market. Unfortunately, this approach does not apply to too many companies. The future demand for each product to be produced must be known before the model can be effective.

Another more common approach is to balance the plant with time. Consider the ABC company. They make over 300 different end items (EIs). They are a mixed job/flow shop. Each EI consists of the same motor, a unique injection molded case, and a variety of metal stampings of which many are common parts to many EIs. The plant consists of four major sections: a punch press shop, an injection molding shop, a transfer line that makes motors, and nine assembly lines. They can produce 15,000 EIs per day.

The plant is balanced with time. The punch press shop runs one shift per day, five days a week. It is scheduled with shop orders based on lot sizes. Since many of the stampings are very inexpensive, often a shop order is for a year's supply.

The injection molding shop has 35 molding machines to make cases. Not all machines can make all cases. Molding is a slower operation. This shop runs two shifts per day, five days per week. It is also scheduled by shop orders based on which cases are required.

The transfer line is the real constraint of the plant. It can produce about 3600 motors per shift. It runs three shifts per day, up to seven days per week. It is occasionally shut down when motor production gets ahead of motor requirements. If demand would go up to 20,000 EIs per day, this line would have to be redesigned or another line would have to be added.

Assembly lines are set up to run a specific EI. They may run a particular EI for a few hours or for many days. Line balancing is performed on a daily basis, since the number of assembly line workers that are available will vary from day to day. The lines run one shift per day, five to six days per week. The rate of the line varies based on the EI being produced.

The ABC company has balanced their plant by scheduling the shifts and days of work for each department. The supporting departments build inventory in anticipation of the needs of the assembly lines. When demand changes, they have often produced inventory for which they now have no immediate use. Their scheduling logic consists of a combination of push and pull logic. The support shops (punch press, injection molding, and motor transfer line) push in anticipation of need. The assembly lines pull from inventory the components that are needed to produce the EIs that are required today. If demand patterns remain reasonably close to what was predicted (forecasted), then this approach works well. If the forecast is only about 60 % accurate, then a lot of inventory will be built that will not have any immediate requirement and therefore incur carrying costs.

Balancing the plant is an effective approach if the demand patterns are known in advance. If the demand patterns can vary, and many products are involved, then an effective high level scheduling system (production planning and master schedule planning) is required to properly manage production plans and inventory levels.

EXTEND THE CASE STUDY

As has been said earlier, the case study example is a very simple example of the mixed environment problem. To fully appreciate the problem, the case study should be extended in both depth and complexity.

For an in-depth extension, the planning at the master schedule level should be adjusted until the plan meets the rough-cut capacity test. This master schedule of planned rates and orders should then be fed to material requirements planning to develop planned shop orders and flow line rate schedules. These plans should then be tested against a set of capacity requirements planning logic.

An ideal method to perform the above steps is with a personal computer using a simulation package. Some of these programs even provide animation so that when you have developed what you consider to be the ultimate production plan, it will show you where queues are building up and where shortages are happening.

We had a lot of constraints on the example. If you have built the above model, you might consider removing those constraints one by one.

In the product structures, change the "required quantity per" to something different than one.

Add input queue times, setup times, output queue times, and move times for all items that flow through a work center. Show move times to and from the flow lines.

When multiple products are produced on one line, assume that some work stations will have to be changed from the production of one product to the next. This represents work station changeover costs that can occur within a family of products (run on the same line). This has to relate to how many of each product you want to ship when. If you make As and Bs, and you only ship once a week, then you can make all the As that you need followed by all the Bs that you need for the week. But let's say that you ship every four hours. You also make As, Bs, Cs Ds, and Es on this line. Since the work station changeover costs will vary from any one product to any other one product, you should determine a least cost build plan such as 50 Cs, followed by 200 Es, followed by 175 As, and so on.

Extend the above approach by assuming that you can build more that one product family on a line. This now requires more than just work station changeover costs. It requires that the entire line be torn down and reassembled. Line changeover costs will vary if a line set up for family D is now to be set up for family B versus family C, or any other family. You plan should reflect the least-cost solution to sequencing multiple families on the line and then, within each family, sequencing the family members.

Adjust the model to accommodate alternate facilities. For example, this component is always built in this work center or on this flow line because it is the most economical way to do it. However, when peak demand periods arise, we can violate the "always" and use another facility or even a subcontractor. These alternate facilities will almost always have different build, move, queue, and setup times.

Consider the utilization of multiple planning horizons based on the planning level. For example, you might have the following horizons for:

Master Production Family Level Planning. A 36-month horizon based primarily on forecasted data. The first 12 months may consist of 12 planning periods (buckets), followed by the next 12 months consisting of four buckets (quarters), followed by the last 12 months consisting of two buckets. Allow hedge inventory to be developed for only 12 months.

Master Schedule Planning. An 18-month horizon consisting of monthly buckets and based on a blend of forecasted demand and actual orders. Up front, add a time fence of one month that says during the first month of the total horizon you will consider only actual customer orders in your plan and not a blend of the forecast and customer orders. Allow hedge inventory to be developed for only six months.

Material Requirements Planning. A six-month horizon in weekly buckets based on the master schedule and adjusted daily based on customer order changes.

Shop Order/Schedule Release. A four-week horizon the first week of which is in daily buckets followed by three weekly buckets. It is vital that this planning level also be adjusted by changes in actual customer orders.

Extend the lead times so that a product (end item) requires more than one day for the production process. Consider taking the low-volume items and having them require five or six days for production. This will provide an interesting impact on your model regarding facility utilization and resource calculations.

Readjust the lot sizes so that some are fixed quantities, some have a multiple of a basic lot (as was done in the example), and some can be anything above a basic lot-size minimum.

YOUR OWN MODEL

If you extend the case study example with the above items, you will lean a lot about scheduling/planning logic as you apply each extension.

You will, however, still have extended a very simple example of a production facility. The true test is to model a live facility. This will involve some analysis on the required production volumes by product. For example, in the above referenced ABC company, the author found that of the 300 end items, 17 of them contributed 97% of the company's profit. Modeling was therefore not done for 300 items. It was done for 17 items (and the company took a good hard look at why they are producing the other 283 items).

We have now completed all discussions on planning logic related to the case study. The following chapters cover some very serious considerations that must be evaluated when migrating from a job shop to a flow shop.

PART SIX

REPETITIVE IMPLEMENTATION CONSIDERATIONS

CHAPTER FOURTEEN

COST REDUCTION AREAS

WHY THE PROBLEM?

Previous chapters address the issue of where high level planning systems fit, and if they do fit, the logic of performing the planning. This issue was covered in depth since many companies consider the potential changes in planning logic to be the prime concern when migrating to a repetitive or Just-In-Time manufacturing environment.

There are, however, a whole different set of issues that must be addressed in addition to systems planning logic for a successful migration to repetitive. These changes apply to how the business and production process areas are managed. They represent a significant change in thinking for the average job shop manufacturer.

The key to success in this new environment is the recognition that "The product selling price is determined by the marketplace and not the producer." That simple statement may seem obvious, but it really is not apparent to many manufacturers. Many producers still use the following equation to determine their selling price.

$$\text{Selling price} = \text{cost} + \text{profit}$$

Let's say that I started producing a product in 1980. My anticipated volume was 50 per month. This product was one more end item in my product line of 450 different items. My cost to produce at this volume in a job shop environment was $20. I wanted to make a 20% profit, so my selling price was set at $24. Since 1980, my volume on this product has increased to 100 per month, my costs have risen to $30 and I have adjusted my selling price to be $36.

A Japanese firm would consider this situation as one with a high potential for repetitive production. They would perform a market analysis and ask, "If we sold this product for $18, how much more of a market could we attract? What if we marketed on an international basis? What would be the impact on product volume if we improved the quality so that the product would be 100% defect free? What if we had significant advertising campaigns?"

Let's say that with all of these considerations, they estimate their potential product volume to be 200 per day. They then perform extensive studies on what the costs would be. They identify suppliers and component delivery frequencies. They identify distribution channels and costs. They analyze potential redesign of the product to simplify the manufacturing process. They estimate the costs in a high volume repetitive environment.

The costs come to $15 per item. The potential profit is therefore $3 ($18 − $15). If this is sufficient profit, they then proceed to put in place a very simple plant with one flow line that has a production capacity of 25 a day, or a quarter of the estimated volume. No leading edge machine tools are installed. This is not an equipment test lab. The supplier and distribution channels are established, and production begins.

If the effort fails, there was a minimum investment risk. If the effort is successful and the demand climbs above 25 per day, another line is added. The following selling price equation is slightly different from the previous one.

$$\text{Selling Price} - \text{Cost} = \text{Profit}$$

The market segment to be addressed determines the selling price. If you cannot produce product with a reasonable profit to meet that selling price, then you would be wise not to try because you will be driven out of business.

The manufacturing industry must recognize that the only way to increase profits is to decrease costs. We will now review some of the potential cost reduction areas.

ELIMINATE OVERPRODUCTION

Do not produce what you cannot sell NOW. This applies to all levels of a product's structure. All too often we see a subassembly with a lot size quantity of 50 and one of the components that goes into the subassembly has a lot size of 500. We have customer orders for 10 of the end item. We are going to build, however, an extra 40 of the subassembly and an extra 490 of the component. This really sounds like poor planning. Why did we have those lot sizes?

Our company started out making a product. As time went on, we added additional products to our production process. We purchased multipurpose machine tools. Based on our product volumes, it appeared to be more economical to buy a machine that would do 20 things than to buy 20 machines that would do one thing each.

The trade off, of course, was that our multipurpose machine required some change over or setup time when we wanted to quit making thing A and start making thing B. The development of our lot size was then based on how much time (cost) was required for a setup, and if we did over produce, what were the carry costs on the excess inventory.

Please note what has just been said. There will be excess inventory. We will pay additional monies to keep it in stock. We are doing this because it takes time to set up the machine. Possibly a more practical approach would be to reduce the setup time.

DECREASE LABOR WAITING

I operate a press. I'm not a setup person, I'm an operator. While my machine is being set up, I drink coffee and chat with fellow workers.

I've now finished my last pallet load of 50 items. The person who does the previous operation has not yet completed the next pallet load. I drink coffee and chat with fellow workers.

My machine is undergoing scheduled (or remedial) maintenance. I am not a maintenance person. While I'm waiting, I drink coffee and chat with fellow workers.

And, of course, while all this waiting is going on I'm getting paid, and adding cost to the product.

If I had a stake in the company; if I viewed it as "my job," instead of "the job I've got right now," would my attitude be different? Would I consider learning how to run different machines, so that when my machine was idle, I could be productive on a different machine? Maybe I could even consider doing minor maintenance on my machines and keeping my own work area clean.

If I did those things, I would have to be concerned about cost reductions. I would be more productive as an operator. I would also be reducing the maintenance and janitorial costs. And, by doing those things, I would be helping to keep my employer competitive in the marketplace and therefore insuring the continuance of "my job."

Note that to change the approaches or current attitudes regarding labor waiting, it must be faced by labor as well as management. An employee who cares must be compensated for caring.

REDUCE TRANSPORTING MATERIAL

Many of our plants today were never designed to produce the products and volumes that they currently produce. They have evolved into their current position.

The worst example that I remember is a plant that was redesigned many times as the product lines changed, but the receiving and shipping docks remained the same. When I last saw them, they had their receiving docks at the end of the production process. As the trucks unloaded, the parts had to be transported the full width of the plant to be put to stores.

More often the problem is in little pieces within the plant. Operation 10 takes place at one location. The next operation, number 20, takes place 40 feet away. The parts weigh 40 pounds each. No one is going to carry each part between operations. What happens is that the parts are stacked on a pallet, and then periodically a fork lift is called to move the pallet from operation 10 to operation 20. It's a simple solution. It is also a process that generates a significant increase in work-in-process inventory. There are 20 parts in queue on this pallet between these two operations. There are also 400 other operations in the plant with similar queues. This has generated a significant amount of work-in-process.

A better approach, if the operations cannot be moved closer together, is to install some type of conveyor system so that the operator could be moving one part at a time as it is completed.

Since any material transportation takes time, the ideal solution is to have both: 1) operations close together and 2) continuous transport facilities. These changes work well in a shop that incorporates a flow line. They are usually difficult to implement in a pure job shop.

INCREASE MACHINE AVAILABILITY TIME

This is sort of like decreasing labor waiting. It amounts to increasing the productivity of the resources that are available, whether they be labor or machines.

Machines do not run because they: 1) have no work, 2) are down for maintenance, or 3) are being set up for the next job.

A machine has a higher probability of remaining scheduled if it is a single purpose machine in a high volume flow line as opposed to a multipurpose machine in a job shop.

Maintenance has always been considered to be a consumer of possibly productive machine time. Some possible solutions are to schedule preventive maintenance during hours when the regular production line is not running, train operators on how to address minor remedial maintenance problems, and require operators to perform the simple oil and grease maintenance functions.

Reduce the time being spent on setting up machines for the next job by simply reducing setups. Some major advances have been made in this area during the last few years with four hour setups being reduced to a few minutes. Of course, the simplest way to reduce a setup is to have a single-purpose machine that never requires a setup.

REDUCE STOCK

The less inventory in stock, the less carry costs (taxes, insurance, building utilities, and so on) that have to be incurred.

From the standpoint of finished goods, this area ties in with the elimination of overproduction. In regard to raw material and components, it also means to not buy a year's supply of an item because "it was a good deal."

But let's say that you did overproduce in the past. What is the status of your situation? Consider this example. Company X had been in business for 20 years. They just hired a new financial vice president who was concerned about inventory costs. The vice president made a simple request: Give me a listing of every item that we have in stock and extend the quantity by our cost of the item. Sequence the list by the date of when the item was last used.

You can probably guess what the listing disclosed. Over $17,000,000 of inventory had no usage during the last five years. They had a carry cost rate of 20%. This meant that every year it was costing $3,400,000 to keep that unrequired inventory in stock.

It is easy to get out of control.

We are going to replace part A with part B. We won't purge item As from stock, because it might be a common part that is still used in other end items. Tomorrow, when we have more time, we'll do a "where used" on item As to verify our suspicion.

I know that we don't use this part in our product any longer, but we cannot remove it from stock because we might need to ship it as a service part.

The shop foreman has come to us (purchasing) three times last month complaining that we are out of stock on part D. Well that won't happen again. I just ordered a year's supply of Ds.

You don't buy a year's supply. You don't even buy a month's supply. You buy (and make) what you need when you need it. Some companies have refined this process to the point where their suppliers ship to them every four hours. Their raw material and component inventories are close to zero. However, due to transportation costs and supplier locations, the more common situation is where a supplier will ship several times a week or weekly.

ELIMINATE WASTE LABOR MOTION

This is an industrial engineering problem. It is also an employee problem as well.

My job on this assembly line is to insert these six screws into a subassembly. I use an air powered screwdriver that I hold in my right hand. The bench alongside of me has the tray of screws. With a shaker feeder, the screws could be fed to within inches of where I use them and then I would not have to reach to the bench six times to complete my task. It is my responsibility as an employee to make my management aware that my task performance might be improved, and the industrial engineer's responsibility to provide the most practical solution.

A typical plant is often filled with examples of wasted labor motion. The dies are located over by the wall. But they are not used over by the wall, they are used here. Each time a die is required, time (unnecessary time) is expended in getting it.

There are two major problems with the solution to wasted labor motion. Many of our smaller companies today consider the hiring of industrial engineers to be one of the things they can do without (until they get bigger). They do not understand that the design engineer (who defines what the product will look like), the manufacturing engineer (who identifies the steps or operations necessary to build the product), and the industrial engineer (who defines/designs the best way to perform the operation), must all work together and have equal importance in the production of a product.

The other problem is one of employee attitude. I get paid $9.25 an hour. I don't really care if I make 500 things a day or 600. In fact, I'll still receive the same amount of money. It's just a job.

This attitude is starting to turn around. More and more companies now have programs in place to motivate employees to become involved.

ELIMINATE PRODUCT DEFECTS

There are two prime considerations for product defect elimination: total product/process design and 100% inspection. Consider these two situations:

Case One. The product is designed by a design engineer, who is a designer, not a manufacturer. The design engineer hasn't been on the shop floor in five years and passes the design (over the wall) to a manufacturing engineer who builds a routing, or sequence of operation steps, of what has to be done to produce the product. There is no industrial engineer. Since it is a given that there might be some flaws in the design or the process, inspection steps are inserted in the process so that every tenth item is inspected.

Case Two. The product is designed in a department that consists of a combination of design, manufacturing, and industrial engineering who work in conjunction with each other. The product, the required operations sequence, and the operations process are defined as a total design unit. Where possible, steps are inserted in the process that will eliminate the need for an inspection step. Where inspections are required, 100% inspection steps are inserted.

Case two will produce a lower-cost defect-free product. When case two meets case one head to head, case one changes the way the business is run or goes out of business.

DECREASE INDIRECT LABOR

If your job does not add value to the product, why are you here? That may sound a little harsh, but the question is valid. And let's not even consider answers like, "Because I'm the president's brother-in-law."

Indirect functions often exist because they have always existed. The business has changed a lot over the last 20 years, but no one ever evaluates how staff functions, for example, should be restructured.

Here's an extreme example. The company is in the $500 million a year sales range. They have 27 vice presidents. The requirement to make vice president is to have worked for the company for 35 years. This was probably a good program when the company was founded. It tended to reduce the amount of personnel turnover. Now the benefit of paying for 27 vice presidents, when four or five would be enough, should be reviewed.

The problem exists at all indirect levels. When a person becomes a first-level manager, they get a private secretary as part of the job. Why? As we increase overhead costs, we increase the overall cost of our product and make ourselves less competitive. Is it more important to have status and a shaky job position (because you might be going out of business), or less status and a secure job?

The Japanese like to structure their plants with no more than three levels of management. I have seen some U.S. plants structured with 10 levels of management, and with all that "control," they take forever to get something done due to the red tape.

As the production processes are reviewed for better ways to do the job, indirect labor should also be reviewed as to how much value it adds to the product.

DECREASE CAPITAL INVESTMENT RISK

We bought a 25 spindle mill. It can do anything. It also cost a small fortune. Because it cost so much, we try to utilize it as much as possible and schedule every job through it. It rapidly becomes the shop bottleneck. Now what? Buy another mill? Replace this one with a bigger and better one? What if our demand goes down? What if technology causes us to change the way we produce our product and we no longer need the mill? It's a big risk.

Or consider this approach. We needed operation 10 done. We bought a simple machine that did what was required for operation 10. For operation 20, we were able to put together our own simple machine. The job requirement or technology changes, and we modify the machines or scrap them out. The investment risk is lower. And remember these simple machines that did only one thing did not require any setup time.

We have discussed some of the potential cost reduction areas, and in the next chapter we review some of the actions that can be taken to support these cost reduction areas.

CHAPTER FIFTEEN

COST REDUCTION ACTIONS

REDUCE SETUP TIME

This action ties in with many of the following actions. The reduction of setup time can be one of the more significant things that can be done to improve productivity and decrease costs.

Many common sense approaches are being employed today for setup time reductions. The molds weigh 200 pounds. Because of their size and weight, they are stored in the warehouse. The old process is as follows:

The previous job is completed.

The previous mold is removed (3 minutes).

The previous mold is returned to stores, and the new mold is retrieved from stores (15 minutes).

The new mold is roughly positioned (8 minutes).

The setup person is notified, and when available, arrives on site (20-minute average).

The setup person performs the fine adjustments and runs some test samples (30 minutes).

The new job starts (total setup time is 76 minutes).

The new process is as follows:

The previous job is going to complete in 10 minutes.

The new mold is retrieved and positioned on a roller bed next to the left side of the machine.

The previous job is completed.

The previous mold is removed by pulling pins, instead of removing bolts, and positioned on a roller bed on the right side of the machine (one-half minute).

The new mold is rolled into position and inserted in place with pins that accomplish both the rough and fine adjustment process (2 minutes).

Test samples are run (2 minutes).

The new job starts.

The old mold is returned to stores (total setup time is 4 1/2 minutes).

Nothing radical was done to achieve this reduction. The changes were all common sense. What did change was the thinking process. Instead of saying, "Let's develop the optimum lot size to match this setup time," the new thought is, "Let's reduce the setup time to the absolute minimum and maybe we won't need a lot size."

BALANCE THE LINE

There are 20 tasks that must be performed on an assembly line. If each of the tasks takes exactly the same amount of time, then the line is balanced. I have never seen this happen. The most that you can hope for is to get the line as close to balanced as possible.

It is now 8:00 a.m. You balanced this line yesterday with 20 tasks to be done on 16 workstations with 16 people. Today, only 14 people came to work. The same task load now has to be rebalanced for 14 workstations.

There are many line balancing programs available for personal computer equipment. The multi-iteration logic is a classical example for the existence of computers. A computer will do a complex line balancing function in minutes, while it would require weeks for industrial engineer to perform the same task manually.

USE ONE-PIECE TRANSFERS

Let's say that I put 50 parts on a pallet and then move them between operations with a fork lift. It takes 50 minutes for work station 40 to process a pallet of parts. The station is now 20 minutes into this pallet. One part is on the machine. There are 29 parts still sitting on the input pallet and 20 parts sitting on the output pallet. We have a work-in-process inventory of 50 parts at this station.

ʼ Other ways to move material might be with a conveyor belt, a roller conveyor, or overhead hooks on a moving chain. One part is on the machine. One part is arriving. One part is leaving. The work-in-process inventory has a significant reduction.

And now, if we design the system so that I cannot put a part on a hook until the previous part has been removed (by the next work station), we have developed a pull system. I cannot make another part until the next work station uses the last part that I made.

Frequently the concept of one-piece transfer is translated to mean minimum quantity transfer. It can mean one tray of 10 pieces, or one tote of four pieces. Regardless of the approach, substantial inventory reductions are usually realized with one piece transfer.

UTILIZE SMALL LOT PRODUCTION

This is similar to decreasing setups, which decreases lot sizes, and utilizing one-piece transfers. Again, build what you can sell. Do not build for stock unless you are building hedge inventory for a future peak demand period.

UTILIZE SIMPLE, INEXPENSIVE, HOMEMADE EQUIPMENT

This topic was addressed in the last chapter under the discussion on decreasing the capital investment risk. If you can build or buy the simple machine that does one thing, you will lower the capital investment risk when the requirement for that one thing goes away.

This action must be applied with good common sense. If you have high assurance that you will have a requirement for an extended period of time, then you might want to spend the additional money to get the best, maintenance-free machine that you can buy.

There are also machines on the market today that have a high degree of flexibility. This does not mean the ability to make part A, resetup and make part B. It means that it can be programmed to perform a given task, and when the requirement for that task ends, it can easily be reprogrammed to perform a different task. Many of the robots on the market today fall into this category.

OPTIMIZE THE PLANT LAYOUT

You are going to build a flow line. You will be using simple, single-purpose machines. You are having your suppliers delivering directly to your production line instead of the warehouse. You are utilizing one-piece transfers between work stations. Do you think that the plant layout will be different than when it was a pure job shop? Of course it will.

The total production process will have to be considered when a new plant layout is defined. Where will suppliers deliver? How will components be moved to the production line? How will parts move between work stations? Modeling and simulation programs are available for various sizes of computer equipment to assist you in the optimization of the facility layout problem.

TRAIN LABOR FOR MULTIPLE OPERATIONS

Some Japanese adjust their employees' pay scale based on the number of machines that they can operate. They do not concern themselves with the issue of, "I am a milling machine operator, because my union card says so." Management instead says, "Learn one machine and earn $X. Learn how to run two machines and earn $X plus $Y. Earn a special status after learning how to operate five totally different machines. Earn special status when you can perform basic maintenance on each of the five types of machines."

You may be thinking that this is not the American world. You are right, but the American world is starting to understand that maintaining the competitive edge is what counts for both the employees and management. Lose that edge and both parties are out looking for jobs.

UTILIZE FLOW TRANSPORT EQUIPMENT

If you move one piece at a time, you normally will not do that with something like a fork lift. You will use some type of continuous-movement transport mechanism. This can be anything from a simple belt to a transfer line where the tote is moved into a work position, the line stops while the work is performed, and then the tote is moved to the next work station to be positioned. There are no queues, lot sizes, machine setups, or pallets of parts.

There is an interesting side effect when these principles are put into practice. The size of the plant shrinks. All that space to store raw materials, to store work-in-process and to store finished goods is no longer required. Often only a third of the existing facilities are required for the new repetitive manufacturing process.

SCHEDULE PREVENTIVE MAINTENANCE

All too often, preventive maintenance is something that we do when we have nothing else to do. After all, it does not add value to the product. That's right, it does not add value to the product, but the lack of preventive maintenance adds cost to the product.

There are two valid approaches to addressing the preventive maintenance scheduling problem. In the first case, incorporate maintenance scheduling logic into the same logic that schedules work on the machines. There are many maintenance scheduling systems on the market that operate in a stand-alone mode. That is, they schedule maintenance on a machine whose workload is scheduled by a totally different system, and the two systems do not talk to each other. The results usually are that the maintenance will be done only if there is no scheduled work on the machine and, therefore, maintenance is seldom done.

The better solution to the first case is to integrate work and maintenance scheduling into the same system. Few of these maintenance approaches exist in actual practice.

In the second case, to avoid a scheduling conflict, do not schedule maintenance and work during the same time period. First shift runs from 8:00 a.m. until 4:00 p.m. Maintenance is scheduled from 4:00 p.m. to 8:00 p.m. Second shift runs from 8:00 p.m. until 4:00 a.m. when maintenance is again scheduled.

Proper preventive maintenance is very important in keeping total costs to a minimum. If you can fix it now for $1, do it, and don't wait until it breaks and will cost $5 to fix.

PROVIDE DEMAND RESPONSIVENESS

The way to satisfy demand on a short delivery cycle has always been to ship it from stock. That means you had to build it to stock. In our repetitive environment, we are trying to reduce stock, and that includes finished goods inventory.

Consider an example where a company makes products A, B, and C. They have contracts to ship 10,000 As, 5000 Bs, and 5000 Cs during the next 20 days. Assuming the build time to be the same for all three products, they could spend the first 10 days making As, followed by five days of Bs, and finally five days of Cs. And, if they shipped only every 20 days, this approach is satisfactory.

But customers do not want to maintain a month's worth of inventory, just like the company does not want to maintain a month's worth of supplier's parts in their own inventory. Customers do not want a shipment once a month; they want shipments daily. The company can no longer spend the first 10 days of the period making only As. Each day they must make 500 As, 250 Bs, and 250 Cs.

To be responsive to customer demands, the repetitive manufacturer must be able to deal with multiple products, produced in smaller quantities, and provided on frequent delivery cycles. Those delivery cycles may even be as frequent as twice a day.

In addition, the requirement for 10,000 As during the next 20 days was probably issued in the form of a blanket purchase order. This means that although customers plan on buying the total quantity, they will often reserve the right to adjust the quantity that they want shipped each day. Based on the sales of the previous day, they inform the producer to ship 475 As, 258 Bs, and 260 Cs tomorrow.

If repetitive producers are to remain competitive, they must have the flexibility built into planning systems. The long range planning is based on a projection of demand, or a forecast. Tomorrow's planning is based on what has to ship.

UTILIZE NONRECURRENT REPAIRS

The days of the quick fix with bailing wire and chewing gum are over. If it breaks, fix it, but fix it so that it never breaks that way again.

Of course, this means that remedial maintenance takes longer than the quick fix, and production is stopped for this longer period of time. But, with the quick fix philosophy, it will break again and again and again. In total, the production will be interrupted for a longer period of time than if the machine was fixed properly the first time.

It is better to miss a schedule today and know that you have a higher probability of making every future schedule than to miss every day's schedule by a little bit.

APPLY INDUSTRIAL ENGINEERING TIME AND MOTION STUDIES

An industrial engineer's job is to find a better way to perform the process. A time and motion study is one tool. How is it done today, and how long does it take?

There are 42 tasks that are performed on the flow line. Task number 18 takes 6.2 minutes. The next longest task takes 5.5 minutes. What is it that the operator does on task 18 that requires 6.2 minutes? If we understand what is done, then maybe we can come up with a way to simplify the task and reduce the time. If we can get the time closer to 5.5 minutes, then maybe we can get the line more in balance.

CONVERT LABOR MOTION TO MACHINE MOTION

Not every job can be converted to total automation by machine. One company that I worked with produced 20,000 units a day of a product. They did it with a totally automated transfer line. Well, almost totally automated. Two thirds of the way down the line (which was over 100 feet long), there were two people performing a task. It consisted of mounting very small metal stampings on tiny posts. It was more cost effective to do this task with people than to try to automate it.

In many cases, however, it is cost effective to convert labor motion to machine motion. The machine does not take sick leave, require overtime pay to run three shifts, take vacation, or go out on strike. Once correctly set up and programmed to make a quality product, the machine does make the same quality product over and over, with consistency.

The consistent quality and an overall cost reduction are usually the main reasons that this action should be considered when practical.

SET UP LINES TO PREVENT THE NEED FOR INSPECTIONS

Our historic approach to the defect problem is as follows. A task is performed at operation 50. Sometimes, it does not come out just right. To make sure that a defective product is not shipped, an inspection step is inserted between operations 50 and 60.

A more common sense approach is to redesign how the task at operation 50 is performed so that no potential for error continues to exist.

The insertion of the inspection step was a quick fix, just like the bailing wire fix on the machine tool. Unfortunately, the inspection station solves the problem, and the problem is therefore never reviewed as to why it continues to exist. It solves the problem, and its adds to the cost. The better approach is to resolve the reason for the problem and eliminate the need to inspect to see if the problem task actually generated a defect.

APPLY 100% INSPECTIONS

Many situations will remain where inspections are necessary. Do you want to sample inspect 10% of the items or do you want to inspect all of them? If you sample inspect, then you are admitting that you are willing to let some defective products be shipped to your customers.

This is no longer a luxury that can be afforded. Product quality and product cost are the two prime considerations for remaining competitive.

MOVE DECISION MAKING CLOSER TO THE SHOP FLOOR

Let's size our plants at 500 employees or less. It will give us greater flexibility on our products as technology changes. It also allows us to reduce the amount of indirect labor. And, we can probably keep management to a maximum of three levels. And as managers, let's listen.

It may come as a shock to some people, but the person doing the job usually has the best ideas for improving how the job is done. Many suggestion programs fail because of the layers of management and red tape that must be gone through for a suggestion to be approved.

For example, I perform two tasks at my work station. I put nuts on some bolts inserted at the previous work station and I insert some screws. I have one air powered device which uses one kind of a bit for a nut and another bit for a screw, so I am constantly changing bits. My suggestion is to provide me with two air powered devices so I no longer have to spend time changing bits.

What should be the approval cycle for a suggestion of that type? Should it go through the "normal approval cycle?" Of course not. It's a good idea. The cost to implement is less than $100. The immediate floor supervisor should be able to evaluate a suggestion of this nature and take the necessary actions that are required.

The message on this last cost reduction action is that the people in the plant usually know best what can be done to improve the operation. As a manager, you have to come up with ways that make it easy for them to convey the information, and make the ideas become reality.

CONCLUSION – HOW TO DO IT

This book has been based on how to address repetitive manufacturing, both from a systems planning standpoint and with approaches to improve operations.

If you are a producer who has a few products, you don't really need very complex, long range, sophisticated planning systems. You do need to seriously consider the cost reduction actions that have been described.

As a producer of a wide variety of many products, if you do not incorporate both the refined planning system approaches and the cost reduction actions, your product lines will be eaten away until only your least profitable items remain and you are out of business.

Job shop producers must recognize that regardless of how long they have produced products, a newer and better way exists, and if they do not adapt, they will no longer be competitive.